Econometric Analysis of Innovation Activities

European University Studies

Europäische Hochschulschriften

Publications Universitaires Européennes

Series V

Economics and Management

Reihe V Série V

Volks- und Betriebswirtschaft

Sciences économiques, gestion d'entreprise

Vol./Bd. 3139

PETER LANG

Frankfurt am Main · Berlin · Bern · Bruxelles · New York · Oxford · Wien

Dubravko Radić

Econometric Analysis of Innovation Activities

PETER LANG
Europäischer Verlag der Wissenschaften

Bibliographic Information published by Die Deutsche Bibliothek
Die Deutsche Bibliothek lists this publication in the Deutsche Nationalbibliografie; detailed bibliographic data is available in the internet at <http://dnb.ddb.de>.

Zugl.: Frankfurt (Main), Univ., Diss., 2004

D 30
ISSN 0531-7339
ISBN 978-3-631-53482-3
US-ISBN 978-0-8204-7673-5

Printed in Germany 1 2 4 5 6 7

www.peterlang.de

In Memory of
Boris Lučić

Preface

*Htio sam da ne smetam nikome, a da poslužim po slaboj svojoj snazi istini kojoj sve znanosti treba da dvore.**
Vladimir Mažuranić

Writing this thesis meant passing through all stages of an innovation process. It started with the desire to create something new, unnoticeably turned into a struggle against all kinds of problems and ended up with a lot of small improvements. But the pure fact that it is finished means that the diffusion phase has started and that it is now left to the 'market' to decide about its acceptance.

I have to thank numerous people. First of all my thesis supervisor, Prof. Dr. Reinhard Hujer, for giving me this great scientific opportunity and providing an environment which was not only 'crazy' for econometrics. He offered me academic freedom but also orientation and inspiration at the right point in time. Thanks also to Prof. Dr. Heinrich Rommelfanger for his prompt readiness to act as the second thesis supervisor. This study was accepted as a doctoral thesis by the Department of Economics at the Johann Wolfgang Goethe University.

PD Dr. Lutz Bellmann and his team at the Institute for Employment Research (Institut für Arbeitsmarkt- und Berufsforschung) were of valuable help with the datasets. Financial support of the German Research Foundation (Deutsche Forschungsgemeinschaft) is also gratefully acknowledged.

I have profited from several discussion with my former and current colleagues Marco Caliendo, Stefan Kokot, Paulo Rodrigues, Stephan Thompsen, Sandra Vuletić and Christopher Zeiss. Thanks to all of you and also to Birgit Andres-Kreiner, our student assistants and to Jan Dahinten for proofreading. It is hardly believable that I was able to finish a single sentence in such a vivid and funny environment.

*Own translation: I did not want to bother anyone, but to conduce with my limited strength to the truth to which all sciences should be devoted. Vladimir Fran Mažuranić, Croatian poet, born in Novi Vindolski in 1859, died 1928 in Berlin.

Innovations often trigger undesired effects. I hope, however, that especially the following people were not affected. First of all my parents. Thank you for your love and for supporting me throughout my entire life. Thanks also to the rest of my family for their encouragement. Although I guess most of them will never read this, it is in fact written for them. Finally, there is one person left which has done more for me than anyone else by giving me so much more than her overwhelming love. Thank you for everything Marija.

Contents

Introduction

This thesis is concerned with the econometric analysis of innovation activities. Innovations are crucial. Not only from an individual firm's perspective but also from an economy-wide viewpoint. The following simple thought experiment proposed by Richard G. Lipsey points to this outstanding role of technological change and innovations in rising economic growth and living standards:

> Imagine freezing technological knowledge at the levels existing in, say, 1900, while continuing to accumulate more 1900-vintage machines and factories and using them to produce more 1900-vintage goods and services, while training more people longer and more thoroughly in the technological knowledge that was available in 1900. It is obvious that today we would have vastly lower living standards than we now enjoy (and pollution would be a massive problem). The contrast is even more striking if the same thought experiment is used to compare today's knowledge, product and process technologies with those that existed at even earlier times. In the long term, new technologies transform peoples' standards of living, their economic, social, and political ways of life, and even their value systems. (Lipsey, 2001, p. 3-4.)

Thinking about innovations is also of considerable importance for Germany. Not a day passes without news about the decreasing innovation propensity and competitiveness of German establishments. It is therefore imperative to learn more about the innovation process and the forces driving it. The first part of this thesis will try to shed some light on two determinants of the innovation process, namely the role of public R&D-subsidies and organizational changes within establishments.

Although the importance of innovations for economic growth is not questioned any longer, no question in economics is as controversially discussed as its impact on employment. Especially in periods characterized by radical technological changes such concerns are 'en vogue' and do not always yield only sophisticated theoretical debates. The famous Luddite riot is such an historical example. Inspired by the charismatic leader Ned Ludd, the Luddites were a group of English workers in the early nineteenth century who destroyed new stocking frames, wool and cotton mills in fear of loosing their jobs. The movement rapidly spread throughout England until the British government finally harshly suppressed the riots in 1813 by executing 13 Luddites.

Even nowadays many fear that the benefits of technological change are unequally distributed among the society and that especially low skilled employees are in the same position as the weavers had been two centuries before. Indeed, the unemployment rate among individuals with no formal education has increased considerably in the last two or three decades. At the same time the employment prospects of high skilled employees with an university degree remained stable despite of an increasing supply. One of the typical suspects in this context is technological change which manifests itself e.g. in increasing investment in information and communication technology. In the second part of this thesis we will pick up this question again and analyze whether technological change is in fact skill biased or not.

This thesis centers around the following two basic questions: What are the determinants and what are the consequences of technological change? In order to give an answer, in the next chapter we will first of all clarify the notion of technological change and innovations by presenting a classification scheme and discussing different measurement concepts. We will also present several determinants of technological change and discuss possible impacts on the labor demand.

As already stated, this thesis is concerned with an econometric analysis of innovations. Therefore in chapter two we will discuss the major problems arising in such an analysis, namely the problem of sample selection and endogeneity, and present some appropriate econometric models to overcome them. This chapter ends with a small Monte Carlo study which highlights under which circumstances sample selection might lead to biased estimation results.

After these theoretical and econometric considerations, part two of this thesis is concerned with the input side of the innovation process. We will start the discussion in chapter three with an examination of the role of the government in the innovation process, i.e. we will answer the question whether public R&D subsidies lead to an increase in the innovation propensity of establishments or not. Various non-parametrical methods, like matching and conditional difference-in-differences estimators, and parametrical methods, namely multivariate and simultaneous Probit models are applied to this aim.

Chapter four considers firm internal factors and analyzes whether the introduction of more flexible and decentralized organizational structures has an innovation enhancing impact or not. We will apply a trivariate Probit and multinomial models and conduct a simulation to see whether the estimated effects are also of any economic importance.

Chapter five addresses the issue of organizational changes which were previously discussed in more depth. We will try to reveal the major forces driving these organizational changes, like investment in information and communication technology, more versatile capital, advances in human capital and changing preferences in favor of more varied jobs. We will model the number of introduced organizational changes using appropriate count data models.

Having analyzed several determinants of innovation activities, part three of this thesis is concerned with the employment consequences of technological change. In chapter six we will test whether technological and organizational changes are skill biased in favor of high skilled employees. Thereby possible sample selection effects are taken into account by the use of mixed simultaneous equation models. Chapter seven extends the analysis to an unique linked employer employee dataset which not only contains detailed establishment information but also data on wages and formal educational attainments of employees.

Chapter eight takes yet another and closer look at the nature of technological change. Using a multiple treatment matching framework we will analyze whether technological change not only discriminates against low skilled but also against older employees and if skill obsolescence is relevant in this context. Chapter nine finally summarizes the major findings of this thesis and gives an outlook on future research.

Part I

Theoretical and Econometric Foundations

Chapter 1

Economics of Innovation

1.1 Introduction, Definition and Classification

*..., daß euer Jubelschrei über irgendeine neue Errungenschaft von einem universalen Entsetzensschrei beantwortet werden könnte.**

Bertolt Brecht

Although a young one, economics of technological change and innovation is now considered as a stand-alone discipline. Its roots can be traced back to the famous work of Joseph Alois Schumpeter, especially his two influencing books: "Theorie der wirtschaftlichen Entwicklung" first published 1912 in Leipzig (English translation "The Theory of Economic Development", Cambridge 1934) and "Capitalism, Socialism and Democracy", New York 1942.[1] Since then it experienced a rapid development covering a wide range of questions today from the sources of technological progress to its impacts on the economy.[2]

We will start this introductory chapter with a more thorough definition of technological change and a brief discussion of some measurement issues. We will then present the most important determinants of the innovation process and summarize existing empirical studies about the employment consequences of technological change. This might help the reader to relate the contributions of this thesis to the ongoing debate about the causes and consequences of innovations.

---

*Taken from "Leben des Galilei" (1955). Own translation: ... today every invention is received with a cry of triumph which soon turns into a cry of fear.

[1]It should be noted that there is also a large business oriented literature which emphasizes the processes of innovations within firms. For an example see Harvard Business Essentials (2003). We will, however, not touch this strand in the following.

[2]For a recent overview see e.g. Cohen (1995a).

To clarify the notion of technological change it might be helpful to look at figure 1.1 which reproduces the trilogy of technological change proposed by Schumpeter: Invention, innovation and diffusion.[3] The process begins with the invention of new ideas. This stage can be influenced by basic R&D efforts within the establishment and is often closely associated with the results of science.[4] In the second step, the innovation stage, new ideas are further developed into marketable products and processes. A useful classification at this point is the division into product and process innovations. The first means the introduction of new products or services while the latter refers to any changes that enable establishments to produce a given output at lower costs. Process innovations can come along with investments in new capital goods, e.g. information and communication technology, or they may reflect an increase in the production efficiency (disembodied process innovation), e.g. organizational changes. Product or process innovations may be new for the own establishment but not for the market (local innovations) or global ones.

Figure 1.1: The Process of Technological Change

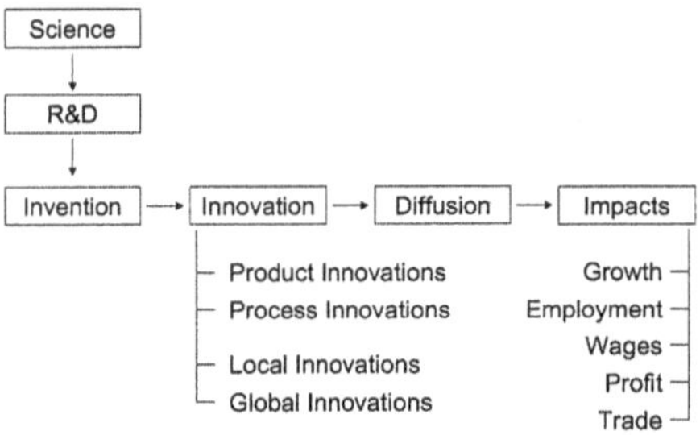

An example for a technologically new product which has not existed before is the first video cassette recorder whereas the substitution of plastic for metal in kitchen equipment represents an improved product.[5] The distinction between product and process innovations is often more blurred since product innovations of one industry might represent a process innovation for another. The introduction of

[3]Cf. Grupp (1997), especially chapter 1.3 or Griliches (1990).

[4]Clearly this description of the process of technological change is highly simplified and does not necessarily bear a close resemblance to the reality. It should, however, only serve as a conceptual framework to facilitate the definition of the various concepts.

[5]See OECD (1997) for a list of other examples. Note that there might be difficulties, especially in services, to classify innovations in only improved and new products.

cellular phones is an example for a product innovation. Road transport companies on the other hand might use cellular phones to reroute their drivers. From their viewpoint the use of cellular phones therefore represents a process innovation.

In the diffusion stage the new products or processes are spread across the economy and adapted by the consumers or other establishments. At this stage the manifold impacts of technological change unfold. Its impact on productivity and economic growth is well documented in the literature.[6] A controversially discussed subject, however, is the relationship between technological change and employment which will be discussed in the last part of this thesis.

The division into invention, innovation and diffusion proposes several measurement concepts of technological change, namely input, byput and output oriented indicators (see also figure 1.2).[7] Input oriented indicators include R&D-expenditures or the number of employees devoted to R&D. According to the *Frascati manual* which outlines the methodology for collecting and using R&D statistics in the OECD countries, R&D "comprises creative work undertaken on a systematic basis in order to increase the stock of knowledge, including knowledge of man, culture and society, and the use of this stock of knowledge to devise new applications" (OECD (1993), paragraph 57).[8] R&D covers basic research, applied research and experimental development. Basic research is "experimental or theoretical work undertaken primarily to acquire new knowledge of the underlying foundations of phenomena and observable facts, without any particular application or use in view" (OECD (1993), paragraph 224). In contrast, applied research aims directly at a certain objective. Experimental development, finally, is directed to "producing new materials, products and devices; to installing new processes, systems and services; or to improving substantially those already produced or installed" (OECD (1993), paragraph 233). Expenditures arising in this context are R&D-expenditures and include labor costs for R&D personnel, current costs for material and equipment, depreciation, capital expenditure, buildings, instruments and equipment.

Although widely used, indicators based on R&D bear several limitations as a measure of technological change.[9] First, they underestimate technological activities in manufacturing and service industries where much of the technical change

[6]See e.g. Griliches (1995) and the literature cited therein.

[7]Cf. Patel and Pavitt (1989).

[8]The Frascati manual is named after the town in Italy where experts from the OECD met in 1963 to discuss definitions and measurement issues of technological change.

[9]Cf. Patel and Pavitt (1989).

Figure 1.2: Measurement Concepts of Technological Change

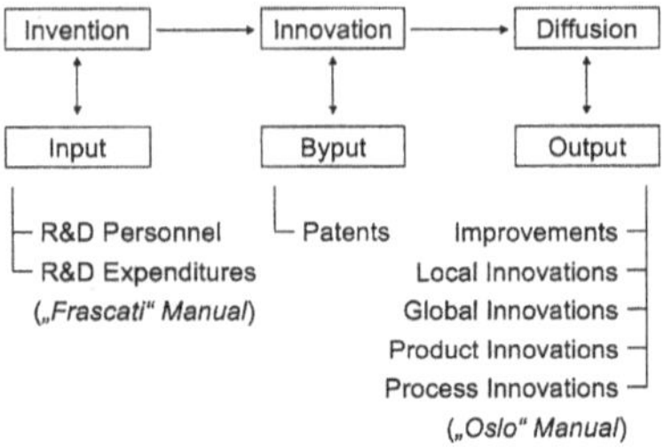

takes place around design and building which is not captured by the concept of R&D. Second, small and medium sized establishments often do not possess a separate business unit devoted to R&D. Using R&D-expenditure or R&D-personnel as a measure will therefore underestimate their innovation activities. And finally, as an input measure R&D-expenditures are only loosely connected with the output of technological change.

Patents granted to establishments are another often used measure for technological activity.[10] Since it is not clear at which stage patents occur and since establishments receive patents for a certain invention but not for an innovation it is reasonable to treat them as an intermediate 'byput' and not as an output oriented measure.[11] Several drawbacks have to be taken into account when considering patents as a measure of technological change.[12] First, not every innovation is based on a patentable invention and not every patent results in a marketable product. Second, in certain industries like the automotive industry, patents play only a minor role as a barrier to imitation and hence differences in patent activities across industries will not reflect differences in innovation activities.[13] Third, institutional and legal factors will also have an impact on patent intensities. And finally, simply looking at patents says nothing about the economic value of the innovation.

[10]Schmookler (1966) pioneered the use of patents.

[11]Grupp (1997) refers to patents as a correlating indicator.

[12]Cf. Griliches (1990).

[13]Cf. Levin et al. (1987). We will return to this point in the next subsection when discussing appropriability conditions as an innovation determinant.

In recent years attempts have been made to measure directly the output of innovations.[14] These efforts culminated in the *Oslo manual* which proposes guidelines for collecting innovation data through large-scale establishment questionnaire surveys.[15] According to the Oslo manual "technological product and process (TPP) innovations comprise implemented technologically new products and processes and significant technological improvements in products and processes. A TPP innovation has been implemented if it has been introduced on the market (product innovation) or used within a production process (process innovation)" (OECD (1997), paragraph 130). Like already mentioned, process innovations can come along with investment in capital goods (e.g. information or communication technology) or refer to organizational changes (e.g. the introduction of total quality management or just in time delivery methods).

The guidelines of the Oslo manual have been implemented in the Community Innovation Survey (CIS) which is an irregular survey of European innovation activities.[16] Additionally, the dataset which will be used in the following empirical applications follows these guidelines, too. We will therefore also rely on output based measures of innovation activities. The advantages of such indicators are obvious since they measure directly the output of the innovation process. There are, however, also disadvantages: Comprehensive establishment surveys are costly and time consuming. Moreover, establishment surveys reflect the subjective view of firms which have to decide what they regard as a product/process improvement or as a product/process innovation.

1.2 Determinants of the Innovation Process

In the following we will summarize the most important determinants of the innovation process, give some theoretical underpinning and present some empirical 'stylized facts'.[17] According to Cohen (1995a) determinants of the innovation process can be classified into firm and industry specific factors. The first include size, market power, cash flow, diversification and technological capabilities whereas the

[14]For other measurement concepts like technometrics and patent citation the reader is referred to Patel and Pavitt (1989).

[15]Cf. OECD (1997). The Oslo manual is named after the location where the constitutional meeting of OECD and Eurostat took place in 1992.

[16]The CIS has been carried out for the first time in 1992 and repeated in 1996 and 2001. For Germany the Centre for European Economic Research is in charge of the CIS. More details can be found on the following web-page: http://www.cordis.lu/en/home.html.

[17]For a more thorough discussion see e.g. Cohen (1995a) and Janz and Licht (2003).

second group consists of the prevailing demand situation, technological opportunities and appropriability conditions. Out of traditional reasons and in honor of Schumpeter, who was one of the first to introduce the notion of innovation into economic theory, firm size and market power are often considered as separate factors (see also table 1.1).

Table 1.1: Determinants of the Innovation Process

Schumpeter factors	Firm size Market concentration
Firm characteristics	Cash flow Diversification Technological capabilities
Industry characteristics	Demand situation Technological opportunities Appropriability conditions

Firm Size

Hypothesis The first Schumpeterian hypothesis claims that innovation activities increase more than proportionally with increasing firm size.

Theoretical Underpinning Several theoretical explanations for the Schumpeterian size hypothesis are offered:[18] Larger firms have better access to capital markets or more internal funds to finance uncertain and risky innovation projects. R&D activities exhibit economies of scale and scope, i.e. larger firms have better opportunities to diversify the risks associated with innovation activities. Fixed costs associated with R&D investments can be distributed over a larger volume of sales. And finally, there might be complementarities between innovation and certain activities, e.g. marketing or planning, which are more pronounced among larger firms. There are, however, also counteracting effects. Larger firms tend to be more bureaucratic and hierarchical which could hinder innovation activities. Associated with an increasing size is also a loss of managerial control of innovation activities. Hence, the impact of firm size on innovations is not clear cut.

[18]See e.g. Freeman and Soete (1997).

Stylized Facts Most of the conducted empirical studies found that an increasing firm size measured by the number of employees increases innovation activities, e.g. expenditures for R&D.[19] However, this increase is only proportional and associated with a decreasing R&D productivity (innovation output as measured by patents per unit R&D).[20] Additionally, considerable differences between industry sectors have been found. Recent studies pointed to a U-shaped relationship where innovation activities first of all fall with increasing size and then rise again.[21] These findings, however, have to be treated with caution since it is unclear whether any firm size effect is in fact generated by some other underlying factors.

Market Concentration

Hypothesis The second part of the Schumpeterian explanation of innovations predicts that a higher market power is positively related to innovation activities.

Theoretical Underpinning Besides intrinsic and nonmonetary aspects like e.g. prestige, the dominant motive for firms to engage in innovation activities is the extra profit induced by innovations. In a market with low entry barriers and a high competition pressure, these extra profits vanish quickly. Thus, the higher the market concentration and entry barriers, the higher the incentives for innovation since it is easier for the innovator to appropriate the returns from innovation. On the other hand, lack of competition may breed inefficient bureaucratic structures and thus discourage innovations. The optimal market structure therefore might be an oligopoly.[22]

Stylized Facts Most of the empirical studies found a positive albeit weak influence of market concentration on innovations.[23] However, this effect is sensitive to the inclusion of other control variables and therefore might simply reflect other industry related characteristics.

[19]See e.g. Villard (1958) or Cohen, Levin, and Mowery (1987).
[20]See e.g. Scherer (1965)
[21]See e.g. Kamien and Schwartz (1985)
[22]For an empirical justification see e.g. Scherer (1967).
[23]See e.g. Horowitz (1962). However, for Germany Felder et. al. (1996) found a negative relation.

Cash Flow

Hypothesis Financial capacity as measured by cash flow is expected to have a positive impact on innovations.

Theoretical Underpinning Due to imperfect capital markets, firms depend on internal funds in order to finance risky and uncertain innovation activities. Larger firms usually have a higher cash flow and therefore are more able to undertake such activities. The cash flow hypothesis is closely related to the size hypothesis. In fact, it rationalizes the first Schumpeterian hypothesis.

Stylized Facts There is mixed empirical evidence about a positive relationship between cash flow and R&D expenditures and that financial constraints can have a negative effect on innovations.[24]

Diversification

Hypothesis Firms with a larger product diversification have also a higher innovation propensity.

Theoretical Underpinning The results of R&D activities are uncertain and hardly predictable. Firms which have a greater scope of products in their portfolio are more likely to exploit some of the uncertain findings of R&D since they simply possess more opportunities to use this new knowledge. This in turn will have an innovation enhancing effect.[25]

Stylized Facts The results on this hypothesis are mixed. Some studies point to the view that diversification has an innovation enhancing impact.[26] However, the results are not robust and likely to be affected by industry specific effects.[27]

Technological Capabilities

Hypothesis Better firm specific technological capabilities will spur innovations.

[24]See e.g. Kraft (1989) or Müller (1967).
[25]Nelson (1959) was the first to introduce this idea into the discussion.
[26]See e.g. Cohen and Klepper (1992).
[27]See e.g. Scherer (1965).

Theoretical Underpinning Firms differ in their specific technological capabilities. These capabilities are reflected e.g. in differences in the qualification structure of the employees, in the internal organization of R&D, manufacturing and marketing and the ways of information processing. Firms with better in-house R&D capabilities will more successfully pursue innovations and also have better 'absorptive' capacities to gain from outside technological opportunities.[28] In fact, technological capability measures a large part of unobservable heterogeneity among firms.[29]

Stylized Facts All empirical approaches in this context face severe empirical problems since technological capabilities are hard to measure. Existing studies have found evidence that the internal organization has an important impact on innovations. Efficient marketing e.g. which pays special attention to consumer needs or collaborations with the external scientific community fosters innovations.[30] Better in-house 'absorptive' R&D capacities will increase the likelihood of such R&D cooperations.[31]

Demand Situation

Hypothesis The larger and faster growing a market, the more innovation activities can be expected.

Theoretical Underpinning Firms respond with their innovation activities to economic incentives, especially to changing demand situations. Large and fast growing markets will therefore increase the return on investment of innovations.[32] This demand pull hypothesis contradicts the technology push view where certain exogenously appearing innovations are assumed to create their own demand.

Stylized Facts Empirical evidence suggests that there are more complex interaction effects between demand conditions and innovation activities and that mutual causation exists. Demand has an impact on incremental innovations.[33]

---

[28] For a more thorough discussion on this subject see e.g. Cohen and Levin (1989).
[29] See also Geroski (1990).
[30] See e.g. Kline and Rosenberg (1986) or Rothwell et. al. (1974).
[31] For Germany see e.g. Becker and Peters (2000).
[32] See e.g. Schmookler (1962).
[33] See e.g. Geroski and Walters (1995).

However, major radical innovations can increase demand growth which then in turn will spur incremental innovations again.[34]

Technological Opportunities

Hypothesis Better technological opportunities prevailing in a certain industry will increase innovations.

Theoretical Underpinning Different industries are characterized by different technological conditions and opportunities. Examples include the maturity of the used technology, the rate of technological advance, the 'closeness' to science and externally generated knowledge through R&D cooperations.[35] If internal 'absorptive' capabilities are available, such external technological opportunities can be exploited for own innovations. The government has an impact on these general technological conditions, e.g. through the use of public R&D subsidies.

Stylized Facts Technological opportunity is a hardly quantifiable concept.[36] Researchers have therefore used several proxy variables, e.g. access to formal and informal technical knowledge or R&D cooperations which all were found to have a positive impact on innovation activities.[37]

Appropriability Conditions

Hypothesis More effective means of appropriability and less extensive spillover effects will increase innovation activities.

Theoretical Underpinning Innovation activities are characterized by spillover effects, i.e. firms must always fear that their inventions will be copied or imitated by competitors. If the appropriation of the benefits is not ensured in some way, the incentives for doing innovations will decrease.[38] Granting patents is one solution to this problem. Others include high costs of imitation, secrecy, first mover advantage or costly complementary instruments such as advertising or customer service. While better appropriability conditions increase individual incentives to innovate,

[34] See Walsh (1984).
[35] See e.g. Cohen and Levin (1989).
[36] For a first attempt see e.g. Scherer (1965) who classified industries into different technology clusters.
[37] For Germany see e.g. Felder et. al. (1996) or Veugelers (1997).
[38] For a theoretical discussion on this subject see e.g. Spence (1984).

spillover effects can have a positive net effect for the economy as a whole since they facilitate the diffusion of new knowledge.

Stylized Facts Several empirical findings have shown that patents only play a major role in industries like the pharmaceutical or chemical and that other instruments are important as well.[39] The net effect of better appropriation on the whole economy is, however, still not clear.

Having presented the most commonly discussed factors of the innovation process, the two most important questions are: Which factors are important and which factors can be influenced? Although empirical evidence suggests that the demand situation, technological opportunities and appropriability conditions are able to explain a considerable part of the variation in firm innovation activities, there are hardly any comprehensive studies which were able to quantify these impacts.[40]

The second question, however, is even more important. Although factors like market concentration, technological opportunities and appropriability conditions will vary over time, at least in the short run they must be assumed as exogenously given. Firm specific factors, like internal organizational structures or governmental interventions on the other hand might be appropriate ways to enhance innovation activities. In the next chapters we will therefore pay special attention to the impact of these factors on the innovation process.

1.3 Technological Change and Employment

The previous considerations should have made clear that technological change is not an exogenously given event but related in a complex way to a variety of determinants. It is, however, not only the endogenous nature of technological change which is still puzzling but also its manifold impacts on the economy. Especially the relationship between technological change and employment is still a controversially discussed topic. In the following we will present the most important

[39] See e.g. Mansfield (1986). For Germany see Felder et. al. (1986).

[40] See Cohen (1995a). Chapter eight of this thesis contains an econometric analysis of innovation decisions like the introduction of product innovations, process innovations and organizational changes. The estimation results suggest that regarding product and process innovations, internal technological capabilities measured e.g. by the existence of a R&D department play the dominant role followed by cash flow considerations and the demand situation.

arguments in this context, present some existing empirical studies and thereby try to shed some light on the following questions: What is the net effect of technological change on the total level of employment? And how are these net benefits distributed, i.e. who are the winners and losers of technological change?

Impact of Technological Change on the Level of Employment

The discussion of the first question, namely the impact of technological change on the total employment level, is dominated by two opposite viewpoints: The notion of 'compensation mechanisms' and the 'labor displacement' orthodoxy. The first is based on classical writers like Adam Smith (1723-1790), Jean-Baptiste Say (1767-1832) and John Ramsay McCulloch (1789-1864) who believed that labor market forces will counteract the initial negative labor market impacts of technological change. On the other hand, the second more pessimistic view of a 'job less' work has been initiated by David Ricardo (1772-1823) in his later work and Karl Marx (1818-1883).[41]

Depending on its nature, technological change has not only initial impacts on the labor demand but also triggers several market compensation mechanisms which are able to intensify or to counterbalance the initial impacts. Much of the debate between the two mentioned orthodoxies centers around the question under which conditions and in what time period the indirect effects outweigh the direct ones, i.e. do labor markets fully adjust to technological shocks and if they do so in what time span?

Let us first of all consider a labor saving process innovation introduced by an establishment. The same output can now be produced with less labor input and hence labor demand will decrease initially. The total amount of decrease depends on the technology used and the substitution relation between labor and capital prevailing in the establishment. However, although the direct employment effects of process innovations will be negative, there are a couple of counterbalancing indirect effects which have to be taken into account.[42]

The first is due to decreasing prices. Employing less workers means also decreasing production costs which in turn might lead to decreasing prices. Decreasing prices will stimulate demand and therefore increase production and employment again. This channel, however, will only work if reduced costs are translated into

[41] For a historical overview see e.g. Petit (1995).
[42] Cf. Spiezia and Vivarelli (2002).

reduced prices, i.e. the more competitive markets are. But even if this is not the case, extra profits experienced by the innovative establishments may be reinvested thereby creating new jobs.

If the innovative establishment increases its production and reduces its prices, other establishments might benefit as well, namely establishments supplying input factors or those which use the cheaper product as an input factor. Thus, although the initial impact of process innovations at the establishment level might be negative, the net effect for the whole economy or even for the industry is not predictable by economic theory without further assumptions.

Let us now turn to product innovations. If the new product is accepted by the market, product innovations will have a positive employment impact at the level of the innovative establishment. Depending on the relation between new and existing products, additional employment effects are to be expected. If the relation is complementary, the new product will induce additional demand for other products and hence the indirect employment effects will be positive. If it is substitutive, however, the net effect is again not clear cut a priori.[43]

Since economic theory offers us no unambiguous answer about the employment effects of technological change, in the following we will turn to existing empirical studies to see if they can contribute more to this question. Similar to theoretical reasoning this approach is also plagued by several problems: The first is the afore mentioned problem of measurement of technological change. Second, employment effects will be different at different economic levels and hence any microeconometric evidence at the establishment level needs to be complemented by macroeconometric studies at the industry or even macro level which are able to reveal indirect employment effects. Another point concerns the endogenous nature of technological change. In the previous section we have pointed to some possible determinants of innovations like market power or demand situation. Clearly, these determinants will also have an impact on labor demand thus leading to possible sample selection effects. Finding an appropriate estimation strategy in this context, however, is not an easy task. We will present some possible methods to this aim in the next chapter.

Table 1.2 on page 25 contains several existing studies about the employment effects of technological change. Since we will use German datasets in the following, we have concentrated on econometric studies for Germany.[44] Most of them

[43]For a thorough discussion on the employment effects of product innovations see Katsoulacos (1986).

[44]For a recent survey on international studies see e.g. Spiezia and Vivarelli (2002). By concentrating

use firm level data from two different datasets (Ifo Innovation Survey and the Mannheim Innovation Panel) thereby covering a time period from 1980 to 1997. The following picture arises: Product innovations have a positive impact on total employment whereas there is mixed evidence about process innovations. In some studies they are negative while in some they are insignificant. Extending the analysis to the sectoral level confirms these findings to some extent although the study of Blechinger et. al. (1998) yields insignificant results. As an upshot we find that the empirical impacts of technological change are manifold and heavily depend not only on the dataset and econometric methods used for the estimation but also on the proxies for technological change: Product innovations in general increase labor demand while process innovations on the other hand seem to be more harmful to total employment.

Impact of Technological Change on the Structure of Labor Demand

Equally important like the quantitative component of technological change are the qualitative impacts on the structure of labor demand, i.e. the question whether all employment groups are equally affected or if there are 'winners' and 'losers' of technological change. Most of the OECD countries have experienced dramatic changes in their structures of wages and employment. The earnings dispersion in the U.S. and U.K. has increased rapidly since the mid-1970s, especially the earnings of skilled, higher educated, more experienced relative to non-productive employees.[45] In contrast, the wage dispersion in Germany has remained fairly stable or even decreased while the unemployment rates, especially among low skilled employees, have increased markedly.[46]

During the same time the supply of high skilled employees also increased in most of the OECD countries. The fact that despite this accelerating supply the wages and employment rates for high skilled also increased, points to the view that demand side factors common to U.S. and Europe must have dominated the relative supply of high skilled employees. Due to differences in labor market institutions, however, these demand side factors manifest themselves in the case of the U.S. and U.K. in an increasing wage inequality and increasing unemployment rates in most of the European countries.

---

on econometric analysis we neglected studies which make use of other empirical approaches. An early example which uses input output analysis is e.g. Scholz et al. (1989).

[45]Cf. Snower (1999).

[46]Cf. Fitzenberger (1999).

The usual demand-side 'suspects' in this context are an increasing international trade, often paraphrased with the notion of globalization, and a skill biased technological change in favor of high skilled employees. Recently, it has been recognized that organizational changes taking place within establishments might also shift the relative demand toward more skilled employees.[47]

The link between an increasing international trade and a worsening of the labor market position of less skilled employees is the Stolper-Samuelson effect.[48] According to this view OECD countries are abundant of high skilled while their trading partners are abundant of low skilled labor. With international trade, OECD countries will increase their exports of skill-intensive and import more less skill-intensive goods. This will cause the relative prices for skill-intensive goods to rise and consequently the relative demand and the relative wages for high skilled employees. High skilled employees will therefore shift from the less skill-intensive to the skill-intensive sector. Additionally, it can be shown that the intra-industrial skill-intensity will decrease.

A first way to test this hypothesis therefore consists in conducting shift-share analyses and decomposing the total increase in high skilled employment into a 'within' and a 'between' component. If international trade is the driving force of an increasing demand for high skilled employees, most of this change should be due to a 'between' shift, whereas in the presence of a skill biased technological change the skill-intensity should increase within all industry sectors. Berman, Bound, and Machin (1998) have shown that for most of the OECD countries during the 1970-1990 period, a widespread within industry substitution toward skilled labor could be found. They conclude that this pattern indicates a pervasive skill biased technological change. A similar analysis for Germany was conducted by Falk and Seim (2001) for the years 1978-1996. They also find that there is a strong empirical evidence for an increasing skill level within all sectors. Chapter seven contains such a shift share analysis for the years 1993 to 1998. The results again lead to a rejection of the international trade hypothesis.

The globalization hypothesis faces also other difficulties: The increasing demand for high skilled employees in skill-abundant countries should be matched by an increasing demand for low skilled employees in countries with abundant low skilled employees. Studies for less industrialized economies have shown, however,

[47]Other explanations are a progressive tertiarization and a growing de-unionization. See e.g. Snower (1999).

[48]See any textbook on international trade, e.g. Markusen et al. (1995)

that the demand for skill has increased in these countries, too.[49] Two other implications of the Stolper-Samuelson effect are an increasing relative price and an increasing relative demand for goods intensive in skilled labor. Studies conducted in this field on the other hand do not find any clear empirical evidence for a change in the price and output structure.[50]

Most economists therefore agree that international trade does not play the most important role in explaining the increasing demand for high skilled employees and therefore turned their attention to another prominent candidate, namely technological change.[51] A first piece of evidence for this hypothesis rests on the undeniable observation that the wage and employment inequality between different skill groups began to increase nearly at the same time when information and communication technology and robotics began to spread.[52] Since then a number of empirical studies has shown that new and more complex technologies also require a higher qualified staff.[53] Additionally, new technologies increase the productivity of high skilled employees more than the productivity of less skilled employees therefore contributing to a complementary relation between new technologies and high skilled employees.[54]

But there are still a lot of other puzzling findings which can not be explained neither the international trade nor by the skill biased technological change hypothesis.[55] Especially the small labor demand elasticities with respect to ICT and the rising inequality within the same skill groups has put forward another explanation for the increasing demand for high skilled employees, namely the organizational revolution that is taking place in many establishments during the last years and which can be characterized as a movement away from a centralized so called 'Tayloristic' organization to a more flexible and decentralized organization. In the second part of this thesis we will therefore put special emphasis on the employment impacts of technological and organizational changes.

Having found some possible explanations for an increasing relative demand

---

[49]For Mexico see the study of Lopez-Acevedo (2002) or for Korea Galhardi (1999).

[50]Cf. Paqué (1998)

[51]For a differing position see Wood (1995).

[52]Cf. Card and DiNardo (2002).

[53]For a survey of international studies see Chennels and Van Reenen (2002). For Germany see e.g. Kaiser (2000) or Falk and Seim (2001).

[54]Cf. Bartel and Lichtenberg (1987). Another explanation for the complementary relation between new technologies and high skilled employees is given by models of bargaining which assume that high skilled workers are more able to capture 'extra rents' from innovation because of higher turnover costs.

[55]Cf. Snower (1999).

for high skilled employees at the establishment level is, however, not sufficient to explain the deterioration of the labor market prospects of low skilled employees. A skill biased technological or organizational change in isolation is not responsible for unemployment or inequality but only in combination with a rigid wage structure and institutional factors. If labor markets were fully flexible, an increasing relative demand for high skilled employees would be compensated by decreasing relative wages for low skilled employees. This situation is illustrated in figure 1.3 by an initial shift in the relative labor demand curve for low skilled employees from $D_0^{L/H}$ to $D_1^{L/H}$ e.g. due to a skill biased technological change (SBTC). Given a stable relative labor supply of low skilled employees ($S_0^{L/H}$), in the new equilibrium E_1 both the relative wage, $w^{L/H}$, and the relative employment for the group of low skilled employees will decrease by the distance DE and AB, respectively.

Figure 1.3: Krugman's Two Sides of the Same Coin-Hypothesis

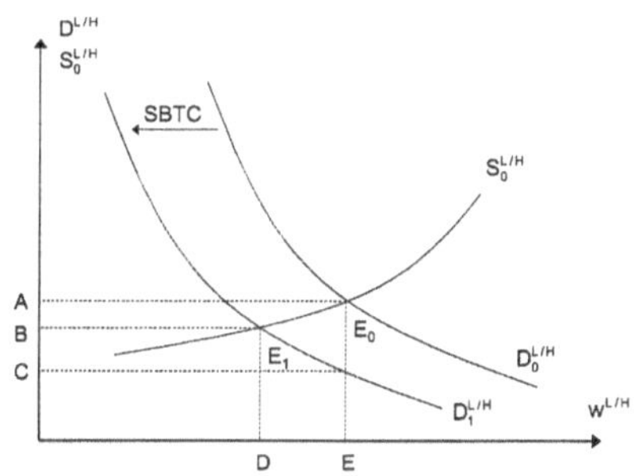

Now consider that some institutional rigidities prevent the relative wages of low skilled employees to adjust to the new equilibrium, a situation which is characteristic for most European countries and caused e.g. by minimum welfare levels amounting to E.[56] As a consequence of this stable wage dispersion unemployment would increase among low skilled employees by AC. In this sense the situation of rising inequality in the U.S. and the situation of rising unemployment in Continental Europe are 'two sides of the same coin' as described by Krugman (1994).[57]

[56]See e.g. Fitzenberger and Franz (1998). Fitzenberger (1999) estimates that the wage dispersion between low and medium skilled employees must increase by 6% on average in order to equalize the unemployment rates for different qualification groups.

[57]In a recent study Puhani (2003) has undertaken an empirical verification of this hypothesis using differentiated data for the U.S., Britain and West Germany. The presented empirical evidence broadly supports Krugman's hypothesis.

The reader should always keep these simple considerations in mind in the following. In the second part of this thesis we will take a closer look at the impacts of technological and organizational changes on labor demand at the establishment level. We will present empirical evidence that these changes indeed discriminate against low skilled and also against older employees. But still, technological and organizational changes *per se* are not responsible for a destabilization of labor markets. Only technological and organizational changes combined with labor market rigidities are part of the cause and hence should not be condemned or even suppressed like the Luddites did nearly two hundred years ago.

Table 1.2: Empirical Studies on the Impact of Technological Change on Total Employment

Study	Dataset	Method	Results
A. Firm level			
Zimmermann (1987)	Ifo (1981-1984)	Cross section model	Negative impact of product and process innovations
Entorf and Pohlmeir (1990)	Ifo (1984)	Simultaneous equation model	Positive (no) effect of product (process) innovations
Zimmermann (1991)	Ifo (1980)	Probit model	Negative effect of process innovations
Ross and Zimmermann (1993)	Ifo (1980)	Probit model	Negative effect of process innovations
Penzkofer (1994)	Ifo (1986-1990)	Simple mean comparison	Positive impact of innovations
König, Buscher, and Licht (1995)	MIP (1993)	OLS and Probit models	Positive (no) effects for product (process) innovations
Rottmann and Ruschinski (1997)	Ifo (1980-1992)	Panel model	Positive (negative) impact of product (process) innovations
Flaig and Rottmann (1999)	Ifo (1968-1995)	Cross section model	Labor saving technological change in the long run
Smolny (1998)	Ifo (1980-1992)	Cross section model	Positive effects of product innovations
Almus, Nerlinger, and Steil (1999)	MIP (1995-1997)	Bivariate Tobit model	Positive impact of innovations
Almus, Engel, and Nerlinger (1999)	MIP (1992-1996)	Bivariate Tobit model	Positive impact of innovations
Blechinger and Pfeiffer (1999)	MIP (1992-1994)	Cross section models	Positive impact of product innovations
Smolny and Schneeweis (1999)	Ifo (1980-1992)	Cross section model	Positive impact of product and process innovations
B. Sectoral level			
Blechinger et al. (1998)	CIS (1992)	Cross sectional model	No significant effect of innovations
Pianta (2000)	CIS (1992)	Cross sectional model	Positive (negative) impact of product (process) innovations

Ifo: Ifo Innovation Survey, MIP: Mannheim Innovation Panel, CIS: Community Innovation Survey.

Chapter 2

Causal Analysis in Econometrics

2.1 The Fundamental Evaluation Problem

*... it was of great importance to me to learn how far the averages were trustworthy.**

Charles Darwin

The previous chapter gave us a first impression about the type of questions which will be addressed in the following. Examples included the impact of organizational changes or public R&D-subsidies on the innovation behavior of establishments or the question whether technological change increases or reduces the demand for labor. These questions have a common structure: The researcher analyzes an outcome variable, e.g. the innovation decision or labor demand, and suspects that a certain treatment like public R&D-subsidies or the introduction of organizational changes has a causal effect on this outcome. The aim of microeconometric evaluation methods consists in testing and quantifying such causal relationships between treatment and outcome. In the following we will briefly discuss some basic econometric issues related to the evaluation of treatment effects and present different estimation strategies which will be used in the following chapters.

The scope of topics for such methods is virtually unlimited and includes questions from medical research ("Is a certain therapy or drug effective or not?"), educational policy ("Does a university degree increase future earnings and by how much?") and business administration ("Does a new distributional channel, e.g. the internet, increase the number of new customers or not?"). Recently, such methods have been successfully applied in evaluating the effectiveness of labor market

*Taken from the introductory remarks to his book on "The Effects of Cross and Self-Fertilisation in the Vegetable Kingdom". Cited after Fisher (1971), p. 28.

programs, e.g. vocational training or public job creation measures.[1] We start with giving a more stringent definition of causality and introduce the following notation to this aim:[2]

i indexes the N individuals in the population, i.e. $i = 1, 2, \ldots, N$.

D_i is the treatment whose effect we want to estimate. Two cases can be distinguished: Binomial and multiple treatments. In the first case we consider the causal effect of one treatment, e.g. receiving a public R&D-subsidy, over no treatment. In this case D_i takes on the value 1 if the i-th individual has received the treatment and 0 if not. In the second case individuals can experience several treatments, e.g. receiving a public R&D-subsidy of type A, B, C or no subsidy at all. If there are M disjunctive outcomes, D_i takes on the values $1, 2, \ldots, M$.

$Y_i(D_i)$ is the outcome variable of interest that we want to attribute to the treatment D_i. By this notation we assume that Y_i is a function of D_i. In the binomial case, $Y_i(1)$ is the outcome of the i-th individual under treatment and $Y_i(0)$ the outcome in case of no treatment. We will introduce the basic concept using the binomial framework in the following. Chapter eight contains a generalization to the multiple treatment case.[3]

The causal effect of treatment for the i-th individual, Δ_i, is the difference between what would have happened if the individual had received the treatment, i.e. $D_i = 1$, and what would have happened if the individual had been exposed to no treatment, i.e. $D_i = 0$. Formally:

$$\Delta_i = Y_i(1) - Y_i(0). \tag{2.1}$$

Equation (2.1) is referred to as the *Roy-Rubin model* of causality and makes clear that thinking about causality requires us to think in terms of counterfactuals.[4]

---

[1] It should be noted that microeconometric evaluation methods only constitute a first step in evaluating all positive and negative effects of a certain treatment and need to be complemented by macroeconomic evaluation methods and cost-benefit analyses which are able to measure the aggregated net-effects. The interested reader can find more details in Hujer and Caliendo (2001).

[2] For a classical discussion on the concept of causality in statistics e.g. Holland (1986).

[3] We will also abstain from sequential treatment models which can be used when analyzing dynamic models. The interested reader can find more details in Sianesi (2003).

[4] Cf. Roy (1951) and Rubin (1974).

The fundamental evaluation problem occurs, because we never observe for the same individual at the same time both $Y_i(1)$ and $Y_i(0)$, but only Y_i which is given by:

$$Y_i = D_i Y_i(1) + (1 - D_i) Y_i(0),$$

i.e. for treated individuals we observe $Y_i(1)$ while for untreated the observable Y_i becomes $Y_i(0)$. The unobservable part is the counterfactual outcome. In this sense the fundamental evaluation problem can be interpreted as a missing data problem and amounts to finding an appropriate approximation for the counterfactual outcome.

There are two solutions to the fundamental evaluation problem depending on the nature of the analyzed question.[5] The 'scientific' solution is applicable when the following assumptions are fulfilled:

- Temporal stability: The value of $Y_i(D_i)$ does not depend on the point of time when the treatment takes place.
- Causal transience: The value of $Y_i(D_i)$ is not affected by previous exposure of the i-th individual to treatment.

Under these two assumptions, the causal effect of D on Y could be simply measured by a sequential exposure of $D_i = 1$ to the i-th individual and then $D_i = 0$, measuring $Y_i(D_i)$ after each exposure. This is an often used approach in natural sciences and especially applicable to physical devices, e.g. electric circuits. An alternative often applicable assumption is the following:

- Unit homogeneity: There are at least two individuals i and j, $i \neq j$, such that $Y_i(D_i) = Y_j(D_j)$ if $D_i = D_j$.

If this assumption is fulfilled the causal effect of D on Y again can be easily measured by exposing the i-th individual to $D_i = 1$ and the j-th individual to $D_j = 0$ and measuring the difference in the outcome-variables.

Both approaches are not applicable to social sciences since neither the assumption of temporal stability, nor causal transience nor unit homogeneity are fulfilled. Individual behavior is rather characterized by its stochastic nature. As a consequence, the individual causal effect D_i is not identified. However, one can focus instead on average causal effects, e.g. the average treatment effect for the entire population which is given by:

$$\Delta^{ATE} \equiv E_i[\Delta_i] = E_i[Y_i(1) - Y_i(0)] = E_i[Y_i(1)] - E_i[Y_i(0)]. \quad (2.2)$$

[5] Cf. Holland (1986).

Δ^{ATE} reflects the expected effect for a randomly picked individual which is exposed to a certain treatment. An implicit assumption necessary to take the average over all individuals is the so called *stable unit treatment value assumption* (SUTVA) which ensures that the treatment effect for one individual is not affected by the treatment effect of the others.[6] SUTVA excludes e.g. cross or general equilibrium effects. We will return to this issue in the next chapter.

The average treatment effect is not the only interesting treatment effect to be estimated.[7] Especially for policy purposes it is not of special interest, because it averages across the entire population which might lead to misleading results. Consider as an example the evaluation of a job training program. In this case the average treatment effect could also include individuals who would never be eligible for participation, e.g. the CEO of an international enterprise as an extreme case.

Thus, it is more interesting to compute the average treatment effect for a subgroup of individuals only. An often used measure is the *average treatment effect on the treated* given by:

$$\begin{aligned} \Delta^{ATE|T} &\equiv E_i[\Delta_i|D_i = 1] \\ &= E_i[Y_i(1) - Y_i(0)|D_i = 1] \\ &= E_i[Y_i(1)|D_i = 1] - E_i[Y_i(0)|D_i = 1] \end{aligned} \tag{2.3}$$

which measures the mean treatment effect for those who actually received the treatment and answers the question whether they were better off than having not received the treatment, i.e. whether the program had any effect at all. The *average treatment effect on the untreated* on the other hand is typically interesting for decisions about extending a program to individuals who were formerly excluded from the program:[8]

$$\begin{aligned} \Delta^{ATE|U} &\equiv E_i[\Delta_i|D_i = 0] \\ &= E_i[Y_i(1) - Y_i(0)|D_i = 0] \\ &= E_i[Y_i(1)|D_i = 0] - E_i[Y_i(0)|D_i = 0]. \end{aligned}$$

Obviously, focusing on expected values does not help us to overcome the fundamental evaluation problem. To see why, look once again at the average treatment effect:

$$\Delta^{ATE} \equiv E_i[\Delta_i] = E_i[Y_i(1) - Y_i(0)] = E_i[Y_i(1)] - E_i[Y_i(0)].$$

[6] Cf. Rubin (1980).

[7] Cf. Heckman, Lalonde, and Smith (1999).

[8] In the next chapter we will discuss the various treatment effects in more detail and give a more thorough justification for their use.

It is important to recognize that what we observe in reality is not $E_i[Y_i(1)]$ and $E_i[Y_i(0)]$ but $E_i[Y_i(1)|D_i = 1]$ and $E_i[Y_i(0)|D_i = 0]$, respectively. And these expectations are not the same things and do not necessarily yield the same results, i.e. in general:

$$(2.4) \qquad E_i[Y_i(1)] \neq E_i[Y_i(1)|D_i = 1] \text{ and } E_i[Y_i(0)] \neq E_i[Y_i(0)|D_i = 0].$$

The same holds for the average treatment effect on the treated defined in (2.3). In this case we can use the average outcome-variable for the treated individuals to approximate $E_i[Y_i(1)|D_i = 1]$ but we have no data at hand for the hypothetical counterfactual outcome $E_i[Y_i(0)|D_i = 1]$. In the following we will discuss this issue in more depth.

2.2 Sample Selection and Endogeneity

Let us return to the inequality in (2.4) and look once again at the core problem which caused this inequality. Assume that Y measures the outcome variable of interest, e.g. whether an establishment has introduced a new product or not, and that D is the binomial treatment variable equal to one if the establishment has received a public R&D-subsidy and zero if not. We suspect that D has a causal effect on Y and aim at measuring this impact Δ, i.e. quantifying the following relation:

$$D \quad \stackrel{\Delta}{\longrightarrow} \quad Y.$$

The major problem with non-experimental data is that assignment into treatment is not random but governed by a sample selection process, i.e. there might be observable, X, as well as unobservable factors, Z, which have a common impact on both the treatment and the outcome decision:

$$\begin{array}{ccc} & X & \\ \swarrow & & \searrow \\ D & \stackrel{\Delta}{\longrightarrow} & Y \\ \nwarrow & & \nearrow \\ & Z & \end{array}$$

As an example, in the next chapter we will present empirical evidence that the group of innovative establishments consists of larger establishments with high qualified employees, R&D-departments, R&D-cooperations and newer equipment.

Looking at the typical establishment which receives public R&D-subsidies on the other hand yields the same structure: Larger establishments with better qualified employees, R&D-departments and newer equipment. In addition to these observable variables, sample selection might also occur on unobservable factors like e.g. management ability which will have an impact on both Y and D. Not taking this mechanism into account will therefore lead to an upward biased estimate of the impact of public R&D-subsidies.

Sample selection occurs when the analyzed relationship between Y and D and the process determining the observability of Y and D are correlated.[9] Thereby different mechanisms can be distinguished (see figure 2.1): If the observability of both variables, Y and D, is governed by the selection mechanism, the resulting sample is said to be truncated.[10] If on the other hand only the observability of Y is affected one obtains a censored sample. The previously discussed application represents a truncated sample whereas one of the first examples for a censored sample comes from Tobin (1958) where he analyzed household expenditures and noted that a large fraction of households reported no purchases. In the following we will concentrate on the truncated sample case only.

Figure 2.1: Classification of Sample Selection Models

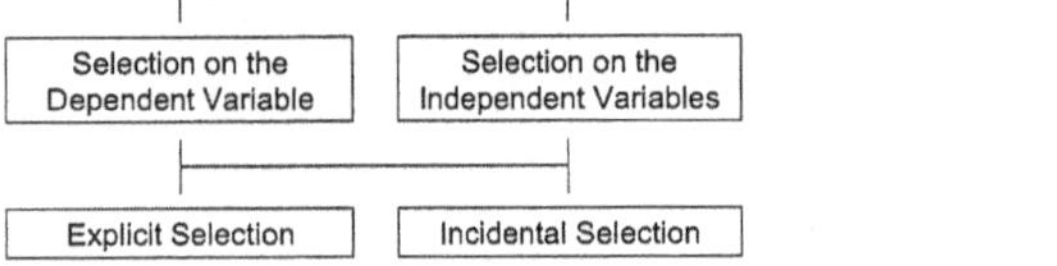

Sample selection might be caused by observable or unobservable factors. Regarding sample selection due to observable variables a distinction between sample

[9] It is important to note that a sample selection rule is not bad per se. But if this rule is connected with the equation of primary interest then it is, because it distorts the estimation results. Consider as an example that you want to analyze the religiousness of the people in your community. Interviewing people leaving a church on Sunday certainly is a selection rule which will yield biased estimation results.

[10] See e.g. Winship and Mare (1992).

selection on measured dependent or independent variables can be made. An example for the first case is the following: Suppose that income is the analyzed dependent variable and individuals are only observed if the income variable is below a pre-specified value. In contrast, selecting only individuals with a specific schooling would be an example for sample selection on independent variables. In both cases selection is a deterministic function of the dependent or independent variables. This relation, however, could also be a stochastic function leading to an incidental selection.

The previously discussed application with innovation as the dependent and R&D-subsidies as the independent variable is an example for an incidental truncated sample selection due to observable and unobservable factors. Figure 2.2 contains a graphical illustration of this case. Assume that treatment D, i.e. receiving a public R&D-subsidy ($D = 1$) or not ($D = 0$), has no impact on the dependent variable Y, i.e. being innovative ($Y = 1$) or not ($Y = 0$). As a result the individual observations denoted by crosses will be scattered uniformly in the DY-diagram.

In the presence of sample selection, however, what we observe differs from this 'true' relation. If especially those establishments which are innovative anyway are also the ones which successfully apply for a R&D-subsidy and vice versa it is more likely that we will only observe individuals marked with bold crosses. Comparing treated with non-treated individuals using such a truncated sample will thus yield an upward biased estimate of the impact of D.

Figure 2.2: Graphical Example for Sample Selection

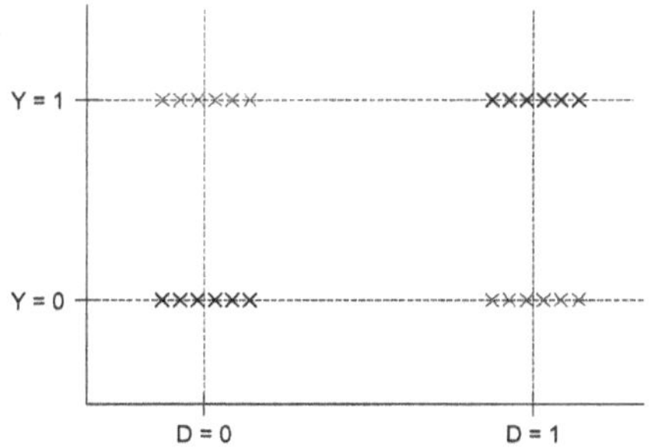

These intuitive arguments can also be formalized.[11] Since both variables Y and D are connected via X and Z, the following independence assumption does not hold:

$$Y(D=0), Y(D=1) \perp D \tag{2.5}$$

which is, however, necessary to approximate any treatment effect. If namely (2.5) is valid, the following two equalities will also hold:

$$\begin{aligned} E_i[Y_i(1)] &= E_i[Y_i(1)|D_i=1] = E_i[Y_i(1)|D_i=0] \\ E_i[Y_i(0)] &= E_i[Y_i(0)|D_i=1] = E_i[Y_i(0)|D_i=0] \end{aligned}$$

and hence the counterfactual outcome for the average treatment effect on the treated, $\Delta^{ATE|T}$, can be estimated using $E_i[Y_i(1)|D_i = 1]$ as an approximation for the otherwise unobservable counterfactual expression $E_i[Y_i(1)|D_i = 0]$. The unconditional average treatment effect can be similarly derived using the observable expression $E_i[Y_i(1)|D_i = 1]$ for $E_i[Y_i(1)]$ and $E_i[Y_i(0)|D_i = 0]$ for $E_i[Y_i(0)]$.

Let us now turn to the question how the concept of sample selection and endogeneity are related. Assume to this aim that Y, D, X and Z are connected via the following linear functions:[12]

$$\begin{aligned} Y &= \Delta D + \beta_1 X + \beta_2 Z + \epsilon \\ D &= \alpha_1 X + \alpha_2 Z + \nu \end{aligned}$$

where X and Z are vectors containing the observable and unobservable factors which have a common impact on both Y and D, respectively.

Now, what would happen if we did not take account of the non-random assignment into treatment and neglected the impact of observable and unobservable variables in the outcome equation? Instead of estimating the 'true' outcome and participation equation we would run the following regression:

$$Y = \Delta D + \epsilon^*.$$

Obviously, the neglected impact of X and Z now would be captured in the residual $\epsilon^* = \epsilon + \beta_1 X + \beta_2 Z$. And since X and Z have also an impact on D, ϵ^* and D would be correlated thus leading to an endogeneity bias.[13] Non-random

[11] Cf. Heckman, Lalonde, and Smith (1999).

[12] The results do not change if we would instead assume non-linear relations. In the following we will conduct a small Monte Carlo study with two Probit equations.

[13] Note that this bias will not disappear even if we include the observable variables X in the equation.

assignment into treatment due to sample selection therefore implies endogeneity of the treatment decision. In the following we will present several estimation strategies which can be used to overcome these problems.

2.3 Microeconometric Evaluation Methods

The previous considerations made clear that the fundamental evaluation problem is in fact a missing data problem. Different estimation strategies differ in the way they approximate the non-observable counterfactual outcome. Thereby the following trade-off arises: If the data stem from an experimental setting, less restrictive identifying assumptions are needed, whereas if they come from a non-experimental setting, it is necessary to impose such restrictions in order to identify any treatment effect. In the following we will introduce the basic ideas and the imposed restrictions of different estimation strategies.[14]

Non-experimental data occur as the byproduct of uncontrolled economic activities whereas experimental data are collected in an experiment where individuals are randomly assigned into a treatment and control group. Field data is data from a natural environment whereas laboratory data comes from an artificial environment. These two characteristics can be combined into four combinations which are all, except for the third case, commonly used in economics and which require different estimation methods.[15]

Environment / Data	Laboratory	Field
Experimental	1.	2.
Non-Experimental	3.	4.

1. Classical laboratory experiments: They offer the advantage that all environmental factors can be controlled for. Additionally, assignment into treatment can be randomized. An often raised objection, however, is lacking realism due to the artificial environment. We will touch this research strand only briefly.
2. Large scale randomized social programs: An example for the US is the National Supported Work Demonstration.[16] Examples for Europe are rare. If

[14]For an overview see e.g. Blundell and Costa Dias (2002) or Vella (1998).

[15]In fact, it makes no sense not to control for the assignment into treatment in a laboratory setting.

[16]The National Supported Work Demonstration was a temporary employment program designed to

certain assumptions are fulfilled, no additional corrections are necessary in order to identify the treatment effect. A special case is the situation of natural experiments where the treatment can be assumed to be exogenous due to external shocks.

4. Non-experimental field data: Datasets prevalent in economics; due to the non-random assignment into treatment econometric models are required which make some identifying assumptions. Since this type of data represents the normal case, we will put special emphasis on appropriate non-experimental empirical models.

2.3.1 Experimental Evaluation Methods

Experiments are one solution to the evaluation problem. The key concept here is the randomized assignment into treatment. Individuals who are eligible e.g. to participate in a social program are randomly assigned into a treatment group which participates in that program and a control group that does not. Random assignment ensures that the potential outcomes are independent of the assignment into treatment, i.e.:

$$Y_i(D_i = 0), Y_i(D_i = 1) \perp D_i$$

and hence the counterfactual outcomes can be approximated using the following equalities which follow from this independence assumption:

$$\begin{aligned} E_i[Y_i(1)] &= E_i[Y_i(1)|D_i = 1] = E_i[Y_i(1)|D_i = 0] \\ E_i[Y_i(0)] &= E_i[Y_i(0)|D_i = 1] = E_i[Y_i(0)|D_i = 0]. \end{aligned}$$

There is a growing number of experiments to evaluate social programs in the US with the most famous one being the National Supported Work Demonstration which was conducted in the mid-1970s. In Europe the number of such large scale experiments is rather limited. The most prominent example is the RESTART experiment in Great Britain which aims at a reduction of youth unemployment.[17]

The advantage of social experiments is that under certain assumptions the parameters of interest are easy to calculate and to communicate to policy makers. Randomization ensures a complete balancing of all relevant observable and unobservable characteristics across treatment and control group. If the sample size

help disadvantaged workers lacking basic job skills to move into the labor market by giving them work experience and counseling them in a sheltered environment. For more details see e.g. Lalonde (1986).

[17]See e.g. Dolton and O'Neill (1996).

is large enough, on average the two groups do not systematically differ except for having received the treatment. As a result, any observed difference in the outcomes of the groups after treatment is solely induced by the treatment itself.

Although some authors like Falk and Fehr (2003), Lalonde (1986) or Ashenfelter and Card (1985) view experiments as the only valid evaluation method, there are several drawbacks which have to be taken into account:[18]

- Ethical issues: It is hardly justifiable to exclude eligible individuals from a program, e.g. programs for disabled or drug-addicted individuals. In this case non-compliance on the part of the institutions which conduct the program is to be expected.
- Randomization bias: A randomization bias occurs when random assignment causes the types of persons participating in a program to differ from the type that would participate in the program without randomization. As a result an unrepresentative sample may occur.
- Substitution bias: Members of an experimental control group gain access to close substitutes for the experimental treatment. In this case the mean outcome-variable of the control group members no longer corresponds to the counterfactual outcome.
- Problem of non-compliance: Members of the treatment group drop out of the program and do not receive the treatment being evaluated. If this attrition process is correlated with the outcome-variable of interest, biased estimates will result.

Experiments restricted to a laboratory environment may mitigate most of the objections mentioned previously but raise other problematic issues. The most important one concerns external validity since laboratory experiments take place in an artificial environment and it might be questionable whether these conditions are also present in reality. A possible subject pool bias is another problematic issue: Experimental subjects are often students which may differ in their behavior from a representative individual. The results may also be questionable because of the small number of participants typically used in laboratory experiments. Finally, subjects may not take their decisions seriously because the financial incentives offered in such experiments are too low.

It is, however, possible to combine the empirical rigor of laboratory experiments with the realism of field data. In a recent study Falk, Lalive, and Zweimüller

[18] Cf. Heckman and Smith (1995).

(2002) present a novel approach to program evaluation. They study the impact of a training program on the likelihood of being invited to a job interview. To this aim unemployed persons who participated in such a program were recruited. In a first step they sent out applications for these persons before they finished the program. In a second step, after they finished the program, exactly the same applications were sent out but now with a course certificate. The impact of the program is measured by the firm's response rate and indicates a negative but insignificant impact.

Sometimes assignment into treatment may also be random due to external shocks which produce exogenous variation in an otherwise endogenous explanatory variable. This situation is referred to as a *natural experiment.*[19] Consider as an example the famous study of Krueger and Card (1994). They used a policy change in wage minimum requirements that affected only one state, namely New Jersey but not the neighbor state Pennsylvania, and examined how employers responded to such an increase in the minimum wage. Another example is the study by Kim and Singal (1993). The outcome variable of interest is the ticket price and the treatment variable the concentration on a particular flight market. Obviously the treatment variable is likely to be endogenous since there might be factors which influence outcome and treatment simultaneously. However, Kim and Singal exploit mergers of certain airlines which affected only markets of these airlines. These mergers can be assumed to happen accidently and hence provide a natural experimental setting.[20]

2.3.2 Non-Experimental Evaluation Methods

Researchers usually do not have access to experimental data. However, econometricians should not complain about this situation, because "it is exactly this so-and-so quality that justifies their legitimacy: If the data were perfect, collected from well designed randomized experiments, there would be hardly room for a separate field of econometrics."[21] In the following we will therefore present appropriate estimation strategies which can be used in case of non-experimental field

[19] Cf. Meyer (1995) or Rosenzweig and Wolpin (2000).

[20] To satisfy curiosity: Contrary to the theory Krueger and Card (1994) found evidence that an increase in the minimum wage increases employment. The results of Kim and Singal (1993) on the other hand showed that prices increased on routes served by the merging firms relative to a group of routes that were unaffected by the mergers. It thus seems that efficiency gains are more than offset by an increasing market power.

[21] Cited after Spanos (1999), p. 28.

data to identify average treatment effects. We start with non-parametrical methods, i.e. methods which make no assumptions about the data generating process for the outcome-variable and how it is related to the sample selection process. The only type of assumptions imposed are various independence assumptions between outcome and treatment. We then turn to parametric models which make explicit assumptions about the data generating process for the outcome- and treatment decision and how both are connected with each other. Figure 2.3 shows the road ahead.

Figure 2.3: Classification of Evaluation Methods

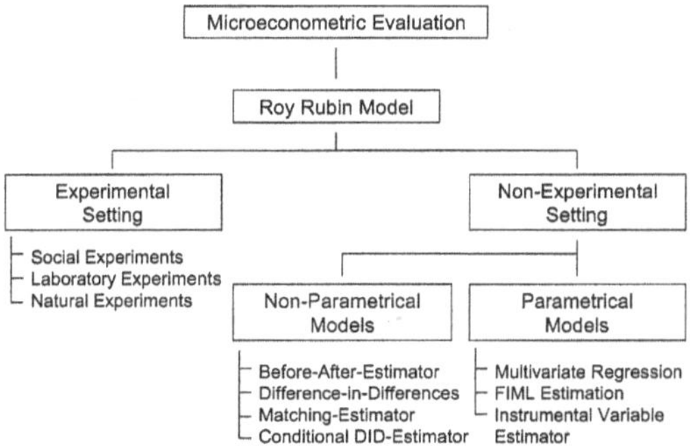

Non-Parametrical Models

The Before-After-Estimator

If longitudinal data are available, a commonly used strategy is to compare a treated individual with him/herself.[22] This strategy is based on the notion that individuals can be in both states at different times. In this case the outcome-variable of treated individuals before the treatment may serve as a proxy for the outcome-variable under no treatment. To incorporate time in the analysis we introduce the following notation:

Y_{it} indicates the post-treatment outcome for the i-th individual

$Y_{it'}$ indicates the pre-treatment outcome for the i-th individual.

[22]Cf. Heckman, Lalonde, and Smith (1999).

The identifying assumption of the before-after-estimator is:

$$E_i[Y_{it}(0)|D_i = 1] = E_i[Y_{it'}(0)|D_i = 1], \tag{2.6}$$

i.e. the counterfactual outcome is approximated by the observable pre-treatment outcome of the treated and the before-after-estimator given by:

$$\widehat{\Delta}^{BAE} = E_i[Y_{it}(1) - Y_{it'}(0)|D_i = 1].$$

If (2.6) holds, the before-after-estimator is equal to the 'true' average treatment effect on the treated:

$$\begin{aligned} \widehat{\Delta}^{BAE} &= E_i[Y_{it}(1)|D_i = 1] - E_i[Y_{it'}(0)|D_i = 1] \\ &= E_i[Y_{it}(1)|D_i = 1] - E_i[Y_{it}(0)|D_i = 1] = \Delta^{ATE|T}. \end{aligned}$$

The before-after-estimator does not necessarily require longitudinal data. As long as the identifying restriction $E_i[Y_{it}(0)|D_i = 1] = E_i[Y_{it'}(0)|D_i = 1]$ holds, repeated cross sections will suffice. As an extension, the before-after-estimator may be calculated for different subgroups of participants, i.e. conditional on X:

$$\Delta^{BAE|X} = E_i[Y_{it}(1) - Y_{it'}(0)|D_i = 1, X].$$

One advantage of this approach is that it only requires information on the participants in a program and their pre-treatment outcome to evaluate the average treatment effect on the treated. A major drawback, however, is the assumption that among participants the outcome in the non-treatment state is the same before and after the treatment, i.e. time-invariant. This assumption excludes changes in the overall state of the economy between t' and t as well as changes in the lifecycle position of participants and is often too restrictive in reality.

Changing economic conditions will also affect the impact of treatments like R&D-subsidies or the effectiveness of labor market programs. Additionally, there might be cases where the treatment has an impact on the pre-treatment outcome of individuals, a phenomenon known as *Ashenfelter's dip*. Orley Ashenfelter evaluated the Manpower Development and Training Act which is a post-schooling training program and noticed that "the earnings of trainees tend to fall, both absolutely and relatively to the comparison group, in the year prior to the training, ... so that some of the observed earnings increase following training may merely be a return to a permanent path of earnings that was temporarily interrupted by one form of transitory labor market phenomenon or another." (Ashenfelter (1978), p. 51.)

One solution to the problem of Ashenfelter's dip consists of using data on the outcome-variable prior to t'. If one has access to enough data on the pre-treatment outcome, $Y_{it'}$ could be even extrapolated using these information. However, this method is only appropriate if the mean outcome evolves as a deterministic function of time.

The Difference-in-Differences-Estimator

The difference-in-differences-estimator can be seen as an extension of the before-after-estimator.[23] Whereas the former compares the outcomes of participants after participation in the program with their outcomes before participation, the latter eliminates common time trends by subtracting the before-after change in non-participant outcomes from the before-after change for participant outcomes.

Pre- and post-treatment information on the outcome-variable for both the treated and control individuals are required. The identifying assumption is given by:

$$E_i[Y_{it}(0) - Y_{it'}(0)|D_i = 1] = E_i[Y_{it}(0) - Y_{it'}(0)|D_i = 0]$$

and states that the mean change in the non-treated outcome is the same for participants and non-participants. Given this assumption the average treatment effect on the treated can be estimated as follows:

$$\widehat{\Delta}^{DID} = E_i[Y_{it}(1) - Y_{it'}(0)|D_i = 1] - E_i[Y_{it}(0) - Y_{it'}(0)|D_i = 0].$$

Under the identifying assumption, one can see that the difference-in-differences-estimator reduces to $\Delta^{ATE|T}$, i.e. the 'true' average treatment effect on the treated:

$$\begin{aligned}
\widehat{\Delta}^{DID} &= E_i[Y_{it}(1) - Y_{it'}(0)|D_i = 1] - E_i[Y_{it}(0) - Y_{it'}(0)|D_i = 0] \\
&= E_i[Y_{it}(1) - Y_{it'}(0)|D_i = 1] - E_i[Y_{it}(0) - Y_{it'}(0)|D_i = 1] \\
&= E_i[Y_{it}(1) - Y_{it}(0)|D_i = 1] \\
&= \Delta^{ATE|T}.
\end{aligned}$$

The difference-in-differences-estimator is based on the assumption of time invariant linear individual selection effects and assumes that these effects are the same on average before and after the treatment. The differences between participants and non-participants then cancel out and eliminate the bias. However, Ashenfelter's dip provides an example where the identifying assumptions of this estimator are violated, i.e. a case where the individual effects are not time invariant.

[23] Cf. Heckman, Lalonde, and Smith (1999).

The Matching Estimator

Matching methods offer a way to estimate the average treatment effect when neither large scale social experiments with a controlled randomization are feasible nor natural experiments which might provide suitable instruments for the estimation.[24] Recall once again that the average treatment effect on the treated is defined as:

$$\Delta^{ATE|T} \equiv E_i[Y_i(1)|D_i = 1] - E_i[Y_i(0)|D_i = 1]$$

and that the fundamental evaluation problem arises since the second term on the right hand side is the counterfactual outcome and thus not observable.

The basic idea of the matching estimator is to approximate this counterfactual outcome by finding in a large group of non-participants those individuals which are similar to the treated individuals in almost all aspects except for the fact that they have not received the treatment.[25] That being done, the difference in the outcome-variables between the participating individuals and the so constructed control group of non-participants is solely attributable to the program. In this sense matching simulates an experiment.

In an experimental setting where assignment into treatment is randomized, the following *independence assumption* applies:

$$Y_i(D_i = 0), Y_i(D_i = 1) \perp D_i$$

thus justifying the use of the mean outcome of the non-treated as an approximation for the counterfactual outcome $E_i[Y_i(0)|D_i = 1]$. If assignment into treatment is not random, this independence assumption is violated.

However, non-random treatment assignment may be random given a set of covariates X.[26] The construction of a valid control group via matching is based on the identifying assumption that conditional on all relevant observable covariates X, the potential outcomes are independent of assignment, i.e. non-participant outcomes have the same distribution that participants would have experienced if they had not received the treatment.

[24] For more details the reader is referred to Heckman, Ichimura, and Todd (1997), Smith and Todd (2000) or Smith and Todd (2000).

[25] Matching estimator are looking for perfect "twins" for the treated individuals. Especially for the US there are a lot of studies which make use of "real" monozygotic twin pairs to estimate various treatments, e.g. the returns to schooling. Examples for such international datasets include: The Australian Twin Registry, the German Observational Study of Adult Twins, the Minnesota Twin Registry, the Osaka University Aged Twin Registry and the Twins' Early Development Study.

[26] Cf. Rubin (1977).

Including all factors which might have an influence on the outcome as well as the participation decision in the vector X, the so called *conditional independence assumption* (CIA) can be written as:

$$Y_i(D_i = 0), Y_i(D_i = 1) \perp D_i | X_i.$$

If the conditional independence assumption is fulfilled it follows:

$$\begin{aligned} E_i[Y_i(0)|D_i = 0, X_i] &= E_i[Y_i(0)|D_i = 1, X_i] = E_i[Y_i(0), X_i] \\ E_i[Y_i(1)|D_i = 0, X_i] &= E_i[Y_i(1)|D_i = 1, X_i] = E_i[Y_i(1), X_i]. \end{aligned}$$

In this case for each given X an unbiased conditional average treatment effect on the treated can be calculated according to:

$$\begin{aligned} \widehat{\Delta}^{MAT|X} &= E_i[Y_i(1)|D_i = 1, X_i] - E_i[Y_i(0)|D_i = 0, X_i] \\ &= E_i[Y_i(1)|D_i = 1, X_i] - E_i[Y_i(0)|D_i = 1, X_i] \\ &= \Delta^{ATE|T,X} \end{aligned}$$

where the second equality follows immediately from the conditional independence assumption. In fact, for the identification of $\Delta^{ATE|T,X}$, it is sufficient to assume that $Y_i(D_i = 0) \perp D_i|X_i$.

In order to obtain an unconditional estimate for the average treatment effect on the treated, we have to assume that:

$$0 < p(D_i = 1|X_i) < 1, \ \forall \ X_i.$$

This assumption ensures that the conditional estimates $\widehat{\Delta}^{MAT|X}$ exist for every X and hence can be averaged over the distribution of X according to:

$$\widehat{\Delta}^{MAT} = E_{X|D_i=1} \left[E_i[Y_i(1)|D_i = 1, X_i] - E_i[Y_i(0)|D_i = 0, X_i]\right] = \Delta^{ATE|T}.$$

Similar to randomization in a classical experiment, the role of matching is to balance the distribution of all relevant pre-treatment characteristics in the treatment and control group, i.e. make these two groups as similar as possible, and thus achieve independence between the potential outcomes and assignment into treatment. There are a lot of practical aspects which have to be addressed in the empirical implementation, e.g. how to construct the control group, the question whether matching is done with or without replacement or how to assess the reliability of the estimation results. The appendix to this chapter contains some of these technical issues. We will also return to such practical aspects in the next chapter where the impact of public R&D-subsidies on innovation activities is evaluated.

The Conditional Difference-in-Differences-Estimator

Matching has become the working horse in the evaluation literature in the meantime. The success of a matching estimator clearly depends on the availability of observable data to construct the set of variables X which satisfy the conditional independence assumption $Y(1), Y(0) \perp D | X$. If one fails to include all observable variables that affect the outcome and the treatment decision simultaneously, a biased estimate will arise. In this sense matching is a data demanding estimator.

Another drawback is the fact that matching assumes the existence of a sample selection process only due to observable covariates. Unobservable time-invariant effects which might be responsible for sample selection as well, like e.g. motivational aspects, are not taken into account. Recall, however, that the difference-in-differences-estimator was able to handle sample selection due to unobservable time-invariant effects. An estimator which combines the advantages of both approaches is the conditional difference-in-differences-estimator.[27]

Recall once again the definition of the difference-in-differences-estimator:

$$\widehat{\Delta}^{DID} = E_i[Y_{it}(1) - Y_{it'}(0) | D_i = 1] - E_i[Y_{it}(0) - Y_{it'}(0) | D_i = 0].$$

The difference-in-differences-estimator assumes that sample selection is solely due to unobservable, linear and time-invariant effects. An extension is the conditional difference-in-differences-estimator which additionally also controls for observables and ensures that only individuals with the same observable characteristics are matched. In particular the conditional difference-in-differences-estimator replaces $E_i[Y_{it}(0) - Y_{it'}(0) | D_i = 0]$ by its matched counterpart. The next chapter contains a detailed description and real case application of this estimation approach.

Parametrical Models

Multivariate Regression Models

We now turn to parametrical methods. We have mentioned previously that sample selection implies endogeneity of the treatment variable. In such a case a simple regression of Y on D will yield a biased estimate of the treatment effect. This bias will arise due to the existence of observable characteristics X which influence Y as well as D. In short, the assumption $Y \perp D$ is likely to be violated. If, however, the conditional independence assumption holds and additionally X is also included

[27] For an application see e.g. Hujer, Caliendo, and Radić (2001).

in the estimation, the impact of D could be estimated free of any bias as it is the case for matching models. In fact, "the differences between regression and matching strategies for the estimation of treatment effects are partly cosmetic" as Angrist (1998) puts it. The essential difference between regression and matching is the weighting scheme used to calculate the average treatment effect at different values of the covariates.

Angrist (1998) has shown that the weights used by the matching estimator are proportional to the probability of treatment at each value of the covariates, whereas in the linear regression model the weights are proportional to the variance of the treatment status conditional on the covariates. He has further shown that the weights used by matching and regression models will be roughly proportional if the values of $p(D_i = 1|X_i)$ are all less than 0.5. On the other hand, if the values of $p(D_i = 1|X_i)$ are all larger than 0.5, the weights are negatively correlated and hence the results of the matching and regression approach might differ. Regression gives more weight to cells in which the proportion of treated and non-treated individuals is similar whereas matching gives more weight to cells in which the proportion of treated individuals is high.

However, a parametrical approach like the linear regression model requires to include the full set of main effects and interaction effects between the covariates X whereas the matching approach is free of such functional form assumptions.

The Full Information Maximum Likelihood Approach

Sample selection may occur due to observable as well as to unobservable factors. If sample selection is solely due to observables, one can either use a parametrical approach like multivariate regression methods or non-parametrical solutions like the matching approach.

But what if there are still differences between the two groups of treated and control individuals after balancing all observable covariates? Or stated differently, what if the conditional independence assumption which is the cornerstone of matching as well as multivariate models does not hold and hence the sample selection mechanism is not only due to observable but also due to unobservable covariates?

As a consequence the treatment decision will still be endogenous and not taking this endogeneity into account might lead to biased estimates. There are two resorts: A non-parametrical approach like the previously mentioned conditional difference-in-differences-estimator or parametrical solutions like the full information maximum likelihood method (FIML).

Recall once again the nature of the problem of sample selection. Assume that Y is the outcome-variable of interest which is related via any function $\mu(\cdot)$ to a set of explanatory variables X and the treatment effect D:

$$Y = \mu(X) + \Delta D + \epsilon. \tag{2.7}$$

Idiosyncratic shocks as well as unobservable factors are captured in the residual ϵ. Now consider the participation decision itself and assume that it is not an exogenous variable but related in a systematic way to a set of variables W and an unobservable component ν according to:

$$D = I(D_i^* = \varphi(W) + \nu \geq 0) \tag{2.8}$$

with $I(\cdot)$ as the indicator function and W potentially capturing some of the variables contained in X as well. Assuming that D is not an exogenous variable is especially reasonable in a non-experimental setting where assignment into treatment is not random. Full information maximum likelihood estimation involves assuming a joint distribution for Y and D and then maximizing the resulting log-likelihood function. Although straightforward to apply, FIML bears several problems. First, it may be simple to assume a data generating process for either Y or D but much more difficult for the joint distribution of Y and D. Second, especially if both variables follow non-linear equations or if there are more than only one endogenous variable the maximization may pose numerical difficulties. We will address this issue in chapter six where we introduce so called mixed simultaneous equation models and also present some simpler two-step estimation procedures.[28]

The Instrumental Variable Approach

A 'shortcut' to FIML is the standard econometric approach to endogeneity, namely the instrumental variable (IV) method. It relies on finding an instrumental variable Z for D which is excluded from the outcome equation, is highly correlated with D but at the same time uncorrelated with ϵ. The treatment effect of D could then be estimated consistently using only the part of variation in D that is associated with Z:

$$\widehat{D}^{IV} = \frac{Cov[Y, Z]}{Cov[D, Z]}.$$

[28] The standard Heckman approach is an example for such a two-step estimator.

A famous example for a clever instrument is a study by Angrist and Krueger (1991). Outcome-variable is wage, explanatory variable years of schooling. Obviously there is a positive sample selection in the sense that individuals with better unobservable abilities will not only have a higher educational attainment but would earn higher wages anyway. Angrist and Krueger use the following instrument for years of schooling: Quarter of birth! Why? Two institutional features have been exploited. First, the fact that you can only enter school when you are six years old by September and second that you must remain in school until age 16. Individuals born in the first quarter of a year therefore have completed 10 school years until the age of 16, while the others completed only 9 years regardless of their abilities. Since quarter of birth can be assumed to be uncorrelated with observable and unobservable abilities, it may be used as a valid instrument.

Another example which, however, makes clear that naturally appealing instruments are hard to find in most of the empirical applications comes from Angrist and Evans (1990). The dependent variable in this case was parent's labor supply and the explanatory variable number of children. Their study uses an ingenious instrument, namely the sex composition of the first two births in families with at least two children, to estimate the effect of additional children on parents' labor supply. It is argued that parents of same sex siblings are more likely to go on to have an additional child. Since sex mix is virtually randomly assigned, a dummy for whether sex of the second child matches the sex of the first child, might be a good instrument for further child bearing.

2.4 A Small Monte Carlo Study

The major task of econometrics consists in detecting causal relations between different variables. This task is, however, counteracted by the previously discussed problem of sample selection. Sample selection occurs when observable and/or unobservable factors not only affect the dependent but also some right hand side variables in the outcome equation. In this case the exogeneity assumption of the covariates is violated and the estimated parameters no longer reflect the 'true' causal effects on the dependent variable.

In the following we will conduct a small Monte Carlo study to get a first idea about the severity and likely magnitude of this bias.[29] Most of the empirical

[29]Hartman (1991) and Stolzenberg and Relles (1991) conducted a similar study but, however, assuming a linear outcome equation.

studies which will presented in the following use microeconomic datasets containing decisions and choices of individuals. Such choices are often not made over a continuum but over a limited number of possibilities and alternatives. Examples include the decision of establishments to introduce new products and services or the introduction of organizational changes, i.e. decisions which will be reflected in discrete variables. For the Monte Carlo Study we will therefore assume that the outcome variable, y_1, follows a Probit equation:

$$Y = I\left(Y^* = \Delta D + \beta_1 X_1 + \gamma_1 Z + \epsilon_1 \geq 0\right) \tag{2.9}$$

where $I(\cdot)$ denotes the indicator function which equals one if the expression within the brackets holds true and zero otherwise. Y is determined by a set of variables collected in the vectors X_1 and Z which are assumed to be truly exogenous. We are especially interested in the parameter Δ which reflects the impact of a possibly endogenous treatment variable D. In one of the next chapters Y e.g. will be the innovation decision of establishments and D the fact whether these establishments have introduced organizational changes or not. Δ can only be estimated unbiased if D is truly exogenous given all other control variables, i.e. if $E(D, \epsilon_1|X_1, Z) = 0$.

In many applications, however, it might be the case that D is still endogenous and that this endogeneity is caused by the fact that besides X_1 and Z there are also other unobservable factors which influence both Y and D and whose impact will be reflected in ϵ_1. In this case the true data generating process (*dgp*) is not given by (2.9) but has to be extended by a behavioral equation for D which is again assumed to follow a Probit model:

$$D = I\left(D^* = \beta_2 X_2 + \gamma_2 Z + \epsilon_2 \geq 0\right). \tag{2.10}$$

The endogenous dummy variable D is influenced by variables X_2 which only affect D but not Y and also by variables Z common to both equations. Variables such as X_2, if available, can be used as possible instruments for the endogenous variable D. The residuals $\epsilon = \begin{pmatrix}\epsilon_1 & \epsilon_2\end{pmatrix}$ are assumed to follow a bivariate normal distribution with zero mean and covariance matrix Σ, i.e. $\epsilon \sim N(0, \Sigma)$. For reasons of identification we have to restrict the variances of ϵ_1 and ϵ_2 to one. Hence $\Sigma = \begin{pmatrix}1 & \rho \\ \rho & 1\end{pmatrix}$ with ρ as the correlation coefficient between ϵ_1 and ϵ_2.[30]

If sample selection is only due to observable variables captured in $\begin{pmatrix}X_1 & Z\end{pmatrix}$, both equations will be independent from each other, i.e. $\rho = 0$, and an unbiased

[30] Equations (2.9) and (2.10) constitute a recursive mixed simultaneous equation model of type I.

estimate of Δ can be obtained by simply estimating (2.9). If, however, sample selection is not only due to observables but also caused by unobservable factors, the assumption $\rho = 0$ is no longer valid. The impact of such factors will be reflected in ϵ_1 and ϵ_2 thus leading to a positive correlation between (2.9) and (2.10), i.e. $\rho \neq 0$. In this case simply estimating (2.9) will yield an estimate $\widehat{\Delta}$ which is plagued by a 'hidden' bias.

In the following we will analyze the magnitude of this bias. In conducting the Monte Carlo analysis we try to add as much realistic assumptions to the underlying *dgp* as possible. A common problem in applied empirical research is that there are no naturally appealing instruments for y_2. In our previously discussed example it is likely that all observable variables affecting the innovation decision will also be correlated with the decision to introduce organizational changes and vice versa. We will therefore assume that $\beta_1 = \beta_2 = 0$, i.e. both outcome and participation decisions are only influenced by the same variables Z.

Regarding the set of variables Z we constructed an artificial dataset by drawing $N = 1,000$ observations from a multivariate normal distribution. In order to add more realism we assumed the same number of exogenous variables as in chapter four where we analyze the impact of organizational changes on innovations and additionally the same means and covariances between the exogenous variables like in the real dataset.

For the 'true' parameter vectors γ_1 and γ_2 we used figures which are typically obtained when running simple Probit models with appropriate innovation and organizational change dummies as the dependent variables:

$$\begin{aligned} \gamma_1' &= \begin{pmatrix} -1.25 & 0.65 & 0.55 & 0.35 & 0.09 & -0.11 & 0.26 & 0.39 & 0.70 \end{pmatrix} \\ \gamma_2' &= \begin{pmatrix} -2.57 & 0.60 & -0.02 & 0.60 & -0.02 & -0.05 & 0.45 & 1.58 & 3.15 \end{pmatrix}. \end{aligned}$$

The 'true' treatment effect of D on Y, finally, was set equal to one, i.e. $\Delta = 1$. Y and D were constructed by adding stochastic disturbances ϵ_1 and ϵ_2 taken from a standard bivariate normal distribution with correlation coefficient ρ to the 'true' latent variables Y^* and D^*. A simple Probit estimation of (2.9) and a full information maximum likelihood (FIML) estimation of (2.9) and (2.10) as a benchmark were conducted. This experiment was repeated 100 times for different values of ρ. Table 2.1 contains the sample averages of the estimated Δ and the corresponding standard errors for the Probit and the FIML estimation, respectively.

Additionally, we also report the average relative deviation of the estimated from the true treatment effect and the root mean square error (RMSE).

Table 2.1: Overall Comparison of the Estimators

	Probit Estimation				FIML Estimation			
ρ	$\widehat{\Delta}$	SE	Rel. Dev.	RMSE	$\widehat{\Delta}$	SE	Rel. Dev.	RMSE
-0.9	-0.59	0.12	-1.59	1.59	0.98	0.16	0.02	0.16
-0.8	-0.35	0.13	-1.35	1.36	1.00	0.20	0.04	0.20
-0.7	-0.17	0.11	-1.17	1.17	0.95	0.26	0.07	0.26
-0.6	0.01	0.09	-0.99	1.00	0.95	0.26	0.07	0.26
-0.5	0.21	0.12	-0.79	0.80	0.99	0.31	0.10	0.31
-0.4	0.34	0.11	-0.66	0.67	0.97	0.36	0.13	0.36
-0.3	0.49	0.12	-0.51	0.52	0.91	0.41	0.17	0.41
-0.2	0.68	0.12	-0.32	0.34	0.94	0.50	0.25	0.50
-0.1	0.85	0.11	-0.15	0.19	0.92	0.59	0.35	0.59
0.0	1.01	0.11	0.01	0.11	0.94	0.59	0.34	0.59
0.1	1.17	0.11	0.17	0.21	0.83	0.59	0.38	0.61
0.2	1.37	0.12	0.37	0.39	0.85	0.61	0.39	0.62
0.3	1.56	0.13	0.56	0.58	0.89	0.65	0.43	0.66
0.4	1.79	0.14	0.79	0.80	1.05	0.68	0.46	0.68
0.5	1.99	0.14	0.99	1.00	0.99	0.69	0.48	0.69
0.6	2.28	0.13	1.28	1.29	1.02	0.66	0.43	0.65
0.7	2.60	0.15	1.60	1.61	0.98	0.61	0.37	0.61
0.8	3.05	0.18	2.05	2.06	1.05	0.60	0.36	0.60
0.9	3.63	0.24	2.63	2.64	1.04	0.49	0.24	0.49

As expected a simple Probit model fails to estimate the true parameter when ρ is large in absolute terms. For $\rho = 0.8$ e.g. the true parameter is two times overestimated. In contrast, the FIML approach is consistent for all ρ but yields larger standard errors. For a moderate correlation between the participation and outcome equation ($|\rho| < 0.3$), the relative deviation of the simple Probit is below 0.5. The performance of the Probit and the FIML approach are also demonstrated in figure 2.4 which contains the kernel density estimations for the two estimators. One can clearly see that although unbiased, the FIML estimator exhibits larger standard errors.

We are facing therefore a trade-off between unbiasedness and efficiency. When confronted with a low correlation between the outcome and participation equation due to unobservable covariates, simply neglecting endogeneity may yield parameter estimates which are closer to the true ones than estimating the outcome and the sample selection equation simultaneously. Figure 2.5 reveals that even for $\rho = 0.5$ the root mean square error of a simple Probit model is comparable to a FIML estimation. Accounting for sample selection in the presence of only modest correlation between outcome and participation can make things even worse.[31] However, with this small Monte Carlo exercise we were able to show that not taking sample selection into account at all can lead to seriously misleading results.

Figure 2.4: Kernel Densities of $\widehat{\Delta}$ for the Probit and FIML Estimation

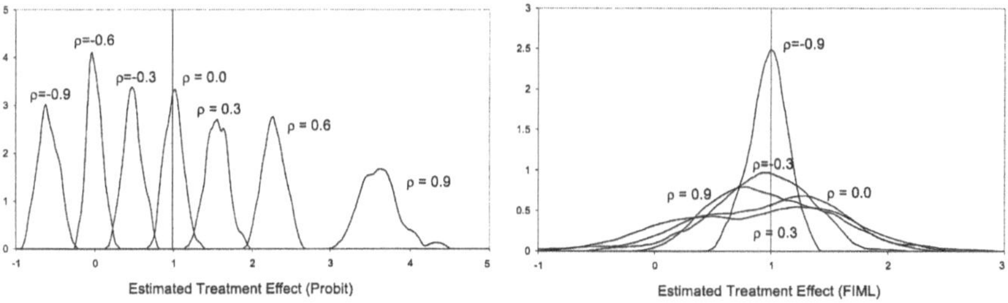

Figure 2.5: Relative Deviation of the Estimated Parameters and Root Means Square Errors

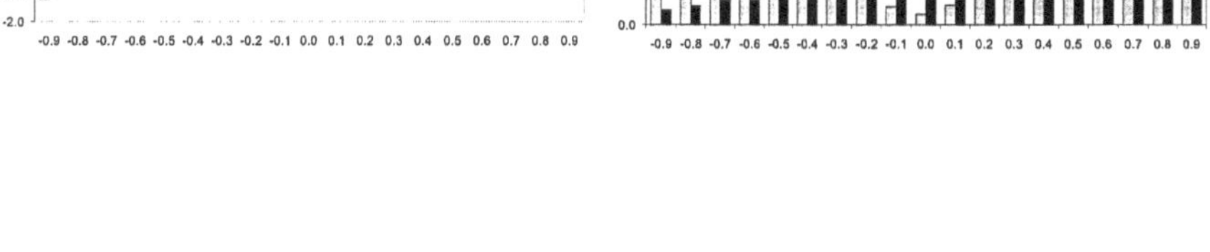

[31]This conclusion is even more valid for limited information methods which will increase the imprecision of the estimates due to their two-step nature. For a similar finding regarding the two-step Heckman correction see Stolzenberg and Relles (1991).

Part II

Determinants of the Innovation Process

Chapter 3

Government: The Role of Public R&D-Policy in the Innovation Process

3.1 Introduction

*Erfolgskontrollen erleichtern den gezielten Abbau von Subventionen.**
German Federal Ministry of Finance

Innovations are crucial, not only from an individual firm perspective but also from an economy-wide viewpoint. However, more than any other economic activity, decisions about innovations and R&D expenditures are plagued by failures of the market mechanism. Innovations represent new knowledge which could be imitated or even stolen by competitors. Furthermore, research as well as the development of new products is a risky and uncertain undertaking and thus must be financed out of own financial resources or venture capital which are both scarce, especially in Germany. As a result of these spillover effects, financial constraints, uncertainties and risk aversion, the level of private innovation activities will be below the social optimum.[1]

All OECD countries are aware of these problems as well as of the importance of technological change and innovations for the future growth.[2] As a response, public

*Bundesministerium der Finanzen (2002). Own translation: Evaluations facilitate a targeted cutback of subsidies.

[1]For a theoretical analysis see e.g. Tirole (1988). Jones and Williams (1998) derived in a growth model a relationship between the social rate of return to R&D and the estimates of empirical literature and conclude that in the U.S. the optimal R&D investment is at least two to four times of actual investment.

[2]In his seminal work, Solow (1957) was able to show, using a growth accounting framework, that in

instruments have been implemented to overcome this dilemma and to stimulate private innovation activities. One of the oldest in this context are patents which were already implemented in Germany in 1877. Additionally there are various other instruments. Some of them, like competition policy or technology transfer, act more indirectly while others, e.g. tax incentive schemes and subsidies, operate in a more direct way to induce innovation activities.

In 2000 total R&D spending in Germany amounted to € 49.8 billions, a considerable part of which was financed by the government. Total public R&D-expenditures amounted to € 15.9 billion with € 2.6 billion paid directly to establishments in the form of R&D-subsidies.[3] The rationale for such measures is to increase the innovation incentives of establishments by lowering marginal costs of R&D and to decrease uncertainties regarding planning reliability. In addition to these direct effects at the establishment level, positive indirect impacts are expected to arise due to spillover effects, e.g. when new technologies and products diffuse and are adopted by other establishments.

However, counteracting effects have to be taken into account as well: At the individual level it could be the case e.g. that establishments would have undertaken innovation activities also in the absence of subsidies or that public R&D-expenditures only crowd out private ones. On a more aggregated level subsidized establishments could rule out non-subsidized ones. The net effect of public R&D policy on innovation activities is thus not clear cut and calls for an empirical evaluation. In order to be thorough and complete, such an evaluation has to address the following questions:

1. Does the subsidy affect a firm's behavior, especially with regard to innovation activities, i.e. what is the causal effect of public subsidies on the individual establishment?
2. Are the individual impacts large enough to yield a social net-benefit or are they counteracted by crowding out effects, i.e. what are the aggregated macroeconomic effects of such subsidies?
3. Is the identified net-benefit the best outcome that could have been achieved for the money spent?

---

the United States e.g. during 1909 and 1949 only a minor part of the per capita growth was induced by an increase in labor and capital but dominated by technological progress.

[3]These figures were taken from the Federal Ministry of Education and Research, BMBF (2002). For an overview of public R&D instruments in Germany see e.g. Czarnitzki et. al. (2003) or Fier and Harhoff (2002).

In the following we will pick up the first question, i.e. estimates the microeconomic effect of policy measures on the innovation activity of German establishments. Thereby we will focus on financial measures, like e.g. subsidies, tax incentives and public credits. Despite the considerable amount of money spent and tight public budgets, empirical evidence, especially for Germany, is rather limited. Table 3.1 contains a synopsis of studies known to us which all point to the view that public R&D subsidies have a positive impact on private R&D and innovation activities.[4] These studies differ with regard to the empirical strategy and outcome variable but all make use of the same dataset, namely the Mannheim Innovation Panel.

Table 3.1: Microeconometric Evaluation Studies of Public R&D Subsidies

Study	Sample	Outcome Variable	Method	Result
Czarnitzki (2001)	East Germany, manufacturing industry	Innovation-intensity	Selection models	Positive
Almus & Czarnitzki (2002)	East Germany, manufacturing industry	R&D-intensity	Matching models	Positive
Czarnitzki & Fier (2002)	Germany, service sector	Innovation-intensity	Matching models	Positive
Licht & Stadler (2003)	Germany, manufacturing industry	R&D-expenditures	Selection models Matching models	Positive

This chapter adds a new piece of evidence to the ongoing political debate about the effectiveness of public R&D measures in the innovation process. An alternative representative dataset, namely the IAB Establishment Panel is used and special attention paid to the problem of sample selection due to observable as well as unobservable covariates. In the next section we will present the dataset as well as some first descriptive results. Section three addresses the problem of sample selection due to observable characteristics by estimating various matching and Probit models. Section four additionally takes unobservable characteristics into

[4]For an overview of international studies see Klette, T. J. and Møen, J. and Griliches, Z. (2000) or David, Hall, and Toole (2000) who surveyed 19 (14) studies on an establishment (sectoral) level from which 10 (12) revealed a complementary relation between public and private R&D expenditures. Irwin and Klenow (1996) analyzed the SEMATECH consortium in the American semiconductor industry while Branstetter and Sakakibara (1998) focused on Japanese research consortia. Lerner (1998) examined the small business innovation research program in the United States while the studies of Griliches and Regev (1999) and Klette and Møen (1999) analyze the situation in Israel and Norway, respectively.

account by employing simultaneous Probit models and conducting a conditional difference in differences estimator. Section five concludes.

3.2 Dataset and First Descriptive Results

The IAB Establishment Panel conducted by the German Federal Employment Office was launched in response to the lack of information about the demand side of the labor market.[5] Its population are all firms employing at least one employee subject to the compulsory social security scheme. The unit of interest is the establishment, i.e. the unit where economic activities take place.[6] All establishments reporting to the German Federal Employment Office are collected in the establishment file from which a stratified representative sample is drawn.[7] The IAB Panel started in 1993 with 4,365 establishments and an average response rate of 71%. After the first wave most of these establishments were reexamined. Additionally, the panel was complemented by first time or repeated registered firms. In 1996 East German establishments were also included in the panel which in 2001 contained about 17,650 establishments.

The panel is organized in a modular form. Some topics are covered annually like changes in the level and structure of employment, questions about employment policy, business volume and investment. Other topics are only covered irregularly, such as information about innovations, for which the latest available data is from 1999/2000. Additionally there is also a special questionnaire about actual topics included into the panel every year.

For our analysis we assume a dependency structure according to figure 3.1: We focus on innovation activities of establishments in 1999 and 2000 and analyze whether they were influenced by public subsidies granted during 1997 and 1998.[8] In terms of the evaluation literature, the innovation decision is the outcome and granted subsidies the treatment variable. Additionally to the treatment variable we also consider a set of covariates from 1997 and 1998 which might have an impact on the outcome variable. The treatment variable itself, i.e. the decision whether

[5]See Hamermesh (1993) and Bellmann (1997).

[6]One advantage of the IAB Establishment Panel is that it includes all establishments and does not impose a minimum requirement regarding employment size or turnover. The Mannheim Innovation Panel e.g. includes only establishments with at least five employees.

[7]The stratification cells are defined by establishment size and industry sector. The selection probability increases with employment size therefore leading to an over-representation of large establishments.

[8]Our dataset only contains information about the innovation decision of establishments referring to the last two years.

or not an establishment has received a subsidy, is assumed to be determined by exogenous variables in 1996.

Figure 3.1: Time and Dependency Structure

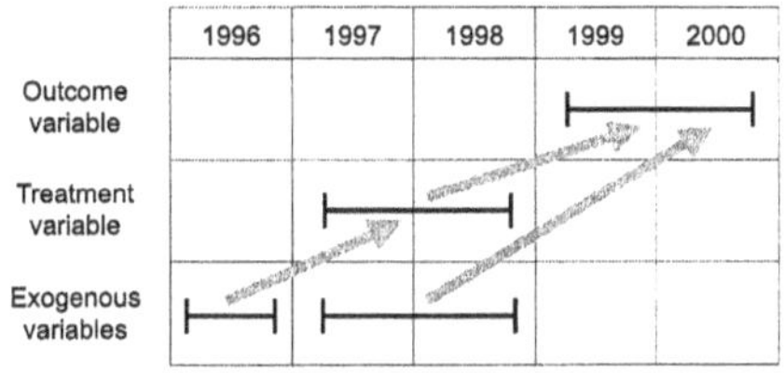

Keeping this time structure in mind, the sample used for the estimation was constructed as follows: In a first step we maintained those establishments which participated continuously in the panel from 1997 until 2001 yielding a sample size of 5,569 observations.[9] In order to isolate the impact of subsidies granted in 1997/1998, we excluded those establishments which received a subsidy in 1999/2000 which reduces our sample size to 3,164 establishments.[10] Finally we excluded establishments from the agricultural and public sector which leaves us with 2,714 observations.

Let us now turn to the precise definition of the treatment variable, i.e. answer the question which public measures are considered in this study. Our dataset contains information if and how many establishments received one of the following subsidies during the years 1997 or 1998:

- Program financed by the federal government and federal states to enhance the regional economic structure ("Gemeinschaftsaufgabe Verbesserung der regionalen Wirtschaftsstruktur") → 35 subsidized establishments
- Programs financed solely by the federal government, e.g. wage subsidies for R&D personnel → 77 subsidized establishments
- Programs financed by federal states to increase the competitiveness of small and medium sized enterprises → 102 subsidized establishments
- Programs financed by the European Union → 55 subsidized establishments

[9]Notice that a questionnaire in year t contains mainly information for year $t-1$.

[10]We were not able to consider public subsidies granted in 1996 because the questionnaire changed between 1996 and 1997 so that the subsidies are not comparable.

- Tax incentives, e.g. investment subsidies or special depreciation → 291 subsidized establishments

- Other programs, e.g. favorable credits from the German Bank for Reconstruction ("KfW") or the European Investment Bank ("EIB") → 90 subsidized establishments.

Establishments which received at least one of these measures are regarded as treated establishments. Although an obvious shortcoming of such an aggregation is that we are not able to disentangle the effects of different measures, several reasons argue for such a proceeding. An obvious one is the necessity to increase the number of observations, e.g. the number of establishments which received federal funds amounts to 77 thus making a reliable estimation of the treatment effect difficult.[11] Another reason concerns the fact that a considerable number of establishments received more than one type of measure so that for these cases a disjunction of the individual effect is not possible.[12] And finally, since the various measures granted by different institutions follow the same objectives, namely to increase the competitiveness and innovation capacity of establishments, it makes sense to focus on the total effect of all public subsidies instead on separate measures.

Table 3.2 contains some basic information about treated and non-treated establishments in the sample. 2,222 (81.87%) of the 2,714 establishments constitute the control group, i.e. establishments which have not received a public subsidy in the years 1997-2001, whereas 492 (18.13%) make up the group of treated establishments. 243 of them (8.95%) received a subsidy only in 1997, 119 (4.38%) only in 1998 and 130 (4.79%) both in 1997 and 1998. In order to increase the total number of observations we pooled the treated establishments in the separate years into one group. Looking at table 3.2 one can see that the participation rate is higher among East German (47.52%) than West German establishments (8.22%). Another feature is the obvious correlation between establishment size and participation rate. The participation rate e.g. for small and medium sized establishments (SME) with 10-250 employees amounts to 24.36% and rises to 32.79% for the group of large establishments with more than 250 employees.[13]

[11] This argument becomes more convincing if one considers that for a reliable estimation we also have to include industrial and regional dummies in the estimation which will additionally reduce degrees of freedom.

[12] In our dataset these were 121 of 492 establishments.

[13] Regarding the definition of small and medium sized establishment we follow the recommendations of the Commission of the European Union from 1996.

Since most of the programs place a special emphasis on SME and East German establishments, all of the following estimations will be conducted separately for these groups.[14]

Table 3.2: Basic Information about the Sample

	Treated	Controls	Participation Rate
West German Establishments	118	1,435	8.22%
East German Establishments	374	787	47.52%
Micro Establishments (Employees < 10)	143	916	15.61%
SME ($10 \leq$ Employees < 250)	229	940	24.36%
Large Establishments ($250 \leq$ Employees)	120	366	32.79%
All Establishments	492	2,222	22.14%

Having defined the treatment variable we now turn to the outcome variable. An often used indicator for innovation activities and therefore the most obvious candidate are expenditures for R&D. As a monetary input variable, R&D expenditures exhibit a close relation to the innovation process especially in knowledge intensive sectors. A disadvantage which was already mentioned in chapter one is the fact that especially smaller establishments do not feature a separate R&D department and thus no explicit R&D expenditures incur. In the case of smaller establishments, innovations are rather generated through practical experience and as a result R&D expenditures would underestimate innovation activities. Another issue concerns the input character of R&D expenditures which do not necessarily reflect the success of the R&D efforts. Since especially new products and services determine the future success and competitiveness of a single firm as well as the economy as a whole, it is more reasonable to focus on an output based measure of the innovation process.[15] In particular we will consider a broad and a narrow innovation concept depending on whether establishments have

- Improved an already existing product/service, introduced an incremental product innovation, i.e. a product which is new for the own establishment but not new for the market, or even a radical product innovation for which a new market had to be created during 1999 and 2000 ($\rightarrow$ broad innovation concept)

[14] See e.g. Czarnitzki (2001) for a discussion of the transition problems in East Germany.

[15] Another more profane reason for choosing this measure is that this is the only one we have at hand!

– Introduced an incremental or even a radical product innovation during 1999 and 2000 (→ narrow innovation concept focusing solely on new products).

A first and tempting way of assessing the impact of public subsidies is to compare the share of innovative establishments in the group of treated and non treated establishments respectively. Table 3.3 contains in the first two rows the appropriate figures. The share of establishments which improved an already existing product/service or even introduced a new one is 42.30% among subsidized but only 36.19% among non-subsidized establishments. The corresponding figures for the narrow innovation concept which focuses only on new products/services and excludes improvements are 27.90% for subsidized and 18.33% for non-subsidized establishments. In both cases the differences are statistically significant which points to the view that subsidies seem to have an innovation enhancing impact.

Table 3.3: Mean Comparison by Treatment Status[a]

Variable	Treated	Controls	p-Value[b]
Broad innovation concept (improvement and new products)	0.42	0.36	0.01
Narrow innovation concept (only new products)	0.28	0.18	0.00
Number of employees	305.68	253.81	0.40
Share of high qualified employees[c]	0.67	0.59	0.00
State of technology (1: Up-to-date, ..., 5: Out-of-date)	2.03	2.12	0.02
R&D department existing	0.22	0.12	0.00
Number of R&D-cooperations	0.56	0.29	0.00
Capital company	0.04	0.07	0.00
Competition intensity (1: None, ..., 4: High)	3.48	3.42	0.20
Market concentration[d]	0.80	0.82	0.00
East German establishments	0.76	0.35	0.00

[a] All variables referring to 1996.
[b] t-test for continuous and test of equality of proportion for dummy variables.
[c] Blue and white collar employees for qualified tasks.
[d] Gini concentration of business volume for 41 different industry sectors.

A simple mean comparison by treatment status, however, will surely not yield an unbiased estimate of the 'true' treatment effect. To see why, let us firstly take a look at the remaining figures in table 3.3 which reveal some of the characteristics of subsidized establishments. In particular, they have a larger share of high qualified employees, are more likely to run a R&D department, cooperate with other R&D institutions like universities, use newer equipment and are more likely located in

East Germany. Now let us turn to table 3.4 which contains the means of the same exogenous variables but now differentiated by the broad innovation concept.

Table 3.4: Mean Comparison by Innovation Status[a]

Variable	Innovators	Non Innovators	p-Value[b]
Number of employees	506.39	112.20	0.00
Share of high qualified employees[c]	0.64	0.58	0.00
State of technology (1: Up-to-date, ..., 5: Out-of-date)	2.02	2.16	0.00
R&D-department existing	0.28	0.05	0.00
Number of R&D-cooperations	0.72	0.11	0.00
Capital company	0.11	0.04	0.00
Competition intensity (1: None, ..., 4: High)	3.56	3.36	0.00
Market concentration[d]	0.81	0.82	0.02
East German establishments	0.35	0.48	0.00

[a] All variables referring to 1996.
[b] t-test for continuous and test of equality of proportion for dummy variables.
[c] Blue and white collar employees for qualified tasks.
[d] Gini concentration of business volume for 41 different industry sectors.

Again we see a familiar structure: Innovative establishments are those with a more qualified staff, R&D departments, more R&D cooperations and a more modern equipment. It thus seems that there is a set of variables which have the same impact on both the innovation decision and the fact whether an establishment has received a subsidy or not. This positive sample selection has to be taken into account out of the following reason: If establishments which have a higher probability to innovate anyway are identical with those establishments which are more likely to receive a public subsidy, a simple mean comparison would yield an upward biased estimate of the true treatment effect. In the following section we will apply appropriate estimation strategies which were introduced in chapter two and which explicitly account for this sample selection mechanism due to observable covariates.

3.3 Sample Selection on Observable Covariates

In this section the sample selection as well as the fundamental evaluation problem which were already discussed in the previous chapter in general will be applied to the application at hand. Let us use the notation introduced in chapter two to this aim. Let D_i be the treatment status of the i-th establishment, i.e. D_i equals one if the establishment has received a public subsidy in 1997/1998 and zero otherwise, and let Y_i be the outcome variable which takes on the value one if the establishment was innovative during 1999/2000 and zero otherwise.

Recall once again that it is our aim to evaluate the causal effect of public subsidies on the innovation capacity of establishments. That is, we want to assess if and to what extent the outcome variable depends on the treatment status. So far we found evidence for a positive correlation between treatment and outcome but also suspected that this is not the true causal effect. Let us thus assume that Y_i is a function of D_i, i.e. $Y_i(D_i)$. The true causal or treatment effect for the i-th establishment, Δ_i, can then be written as

$$\Delta_i = Y_i(1) - Y_i(0), \tag{3.1}$$

i.e. the difference between the outcome of an establishment which received a subsidy, $Y_i(1)$, and the counterfactual outcome under no subsidy, $Y_i(0)$. At this point the familiar fundamental evaluation problem occurs which consists in the fact that the outcome variable for one establishment is given by $Y_i = Y_i(0) + [Y_i(1) - Y_i(0)]D_i$ so that we can only observe either $Y_i(1)$ or $Y_i(0)$ while the other value represents the counterfactual outcome. $Y_i(0)$ is unobservable for subsidized and $Y_i(1)$ for non-subsidized establishments. Hence, the individual treatment effect Δ_i is not identified and we have to turn to the average effect for the entire population or a sub-sample by taking the expectation of (3.1):

$$E(\Delta_i) = E[Y_i(1) - Y_i(0)] = E[Y_i(1)] - E[Y_i(0)] = \Delta^{ATE}. \tag{3.2}$$

Equation (3.2) gives the average treatment effect for the entire population, i.e. in our case it answers the question which impact a subsidy has on a randomly picked establishment. In practice, however, most studies do not focus on Δ^{ATE} but instead condition on participation and estimate the so called treatment effect on the treated given by

$$\begin{aligned} E(\Delta_i|D_i = 1) &= E[Y_i(1) - Y_i(0)|D_i = 1] \\ &= E[Y_i(1)|D_i = 1] - E[Y_i(0)|D_i = 1] \\ &= \Delta^{ATE|D}. \end{aligned} \tag{3.3}$$

$\Delta^{ATE|D}$ estimates the effect of subsidies for those who actually received such a measure, i.e. the mean gain for participants. In the appendix we will give a more thorough justification for its use in our context. Looking at equation (3.3), one can see that we are still left with the fundamental evaluation problem since $E[Y_i(0)|D_i = 1]$ represents the expected value of the counterfactual outcome of a subsidized establishment under the hypothetical situation of no subsidy which is not observable. Solving the fundamental evaluation problem can thus be interpreted as a missing variable problem and amounts to finding an appropriate approximation for this counterfactual outcome.

A simple mean comparison by treatment status, which we conducted previously:

(3.4)
$$E[Y_i(1)|D_i = 1] - E[Y_i(0)|D_i = 0] = \Delta^{ATE|D} + E[Y_i(0)|D_i = 1] - E[Y_i(0)|D_i = 0]$$

will only yield an unbiased estimate of $\Delta^{ATE|D}$ if assignment to treatment and potential outcomes of non-treated establishments are independent, i.e. if treatment assignment is ignorable or formally $Y(0) \perp D$.[16] In this case the bias term $E[Y_i(0)|D_i = 1] - E[Y_i(0)|D_i = 0]$ will vanish and hence $E[Y_i(1)|D_i = 1] - E[Y_i(0)|D_i = 0] = \Delta^{ATE|D}$.

This independence assumption is unlikely to hold outside a non-experimental setting but is more likely to be fulfilled in our application if we additionally take a set of covariates, X, into account. If the following conditional independence assumption holds:[17]

$$Y(1), Y(0) \perp D|X, \tag{3.5}$$

i.e. if the sample selection is solely due to observable covariates, then $E[Y_i(0)|X, D_i = 0] = E[Y_i(0)|X, D_i = 1]$ and the average treatment effect on the treated can be estimated by taking the expectation of the following expression over the distribution of $X|D_i = 1$:

$$\Delta^{ATE|D} = E_{X|D_i=1}\left\{E[Y_i(1)|X, D_i = 1] - E[Y_i(0)|X, D_i = 1]\right\} \tag{3.6}$$

As Angrist (1998) puts it, $\Delta^{ATE|D}$ is obtained as a weighted average of contrasts between the outcome variable of subsidized and non-subsidized establishments at each value of X. Such a balanced comparison controls for all observable differences between subsidized and non-subsidized establishments.

[16]See Rubin (1979).

[17]This assumption is even stronger than necessary. In order to identify the average treatment effect on the treated, $Y(0) \perp D|X$ would be sufficient.

3.3.1 Estimation Strategies

One possible estimation strategy based on (3.6), which was also discussed in more depth in the appendix to chapter two, is to stratify the dataset into cells defined by combinations of each value of X, to calculate within each cell, i.e. conditional on X, the difference between the means of the outcome variables for treated and non-treated establishments and, finally, to average these differences over the distribution of X. Under the assumption that the selection process is solely due to observable covariates captured in X, this exact cell matching will yield an unbiased estimate of the 'true' treatment effect on the treated.

More intuitively, a matching estimator tries to approximate the counterfactual outcome by finding in a large group of non-participants those 'twin'-establishments which are similar to the participating ones in all aspects except for the fact that they have not received a public subsidy. That being done and if we can assume that the selection process is only due to observable covariates, the difference in the outcome variable between participating and matched non-participating establishments is solely attributable to the program. In that sense matching estimators simulate an experimental setting.

A practical obstacle of such a proceeding lies in the so called dimensionality problem which was already introduced the appendix to chapter two. If our vector of covariates X e.g. contains K binary variables, the number of cells amounts to 2^K. If there are also continuous variables like it is the case in our dataset, it becomes even more likely that in some cells there will be only treated or untreated establishments and hence a mean comparison for these cells cannot be conducted.

As a resort, Rosenbaum and Rubin (1983) proposed the concept of propensity score matching. Defining the propensity score as the conditional probability to participate, i.e. $p(X) = p(D = 1|X)$, Rosenbaum and Rubin showed that

$$Y(1), Y(0) \perp D|X \Rightarrow Y(1), Y(0) \perp D|p(X), \tag{3.7}$$

and hence one needs not condition on all covariates contained in X but only on the propensity score $p(X)$. Given this implication, a variety of matching protocols can be used to estimate the average treatment effect on the treated. The appendix to chapter two contains more details to these protocols which all, however, have the following steps in common:[18]

1. Estimate the propensity score using e.g. a Probit or Logit model.

[18]See e.g. Heckman, Ichimura, and Todd (1997).

2. Match treated establishments with those non-treated establishments which are similar regarding the propensity score.

3. Calculate the mean of the outcome variable for treated and matched control establishments. The difference in the means may serve as an estimator of the average treatment effect on the treated.

Some comments might be helpful regarding the first and second step. The quality of the matching estimator heavily depends on the quality of the estimation of the propensity score and whether the so called balancing property is fulfilled. The balancing property states that for a given $p(X)$, treated and control establishments should have the same distribution of X. If this is the case, the differences in the outcome variables are solely attributable to participation. If, however, treated and matched controls are still different in terms of observable covariates after the matching, the estimated treatment effect on the treated may not be reliable.

A first way to improve the matching quality is to impose the so called common support restriction, i.e. drop those controls that have a propensity score lower than the minimum or higher than the maximum propensity score of the treated establishments and vice versa in order to ensure that the matched pairs are comparable. Another way is matching not only on the propensity score but additionally on some covariates, like e.g. industrial sector.

Step two concerns the question how to construct a control group, i.e. how to define similarity between establishments. To this aim several procedures for matching on the propensity score have been suggested. In its most general formulation a matching estimator is given by

$$\Delta^{Matching} = E\left\{ [Y_i(1)|D_i = 1] - \sum_{j \in \{D_j=0\}} w(i,j)[Y_j(0)|D_j = 0] \right\}, \tag{3.8}$$

where $w(i, j)$ represents the weight attached to the j-th non participating establishment. The more similar the i-th and j-th establishments are regarding their propensity score, the higher the weight. Different matching estimators differ in the weights attached to non-participating establishments.

Nearest neighbor matching e.g. requires to choose for every participating establishment that out of the group of non-participating which is closest regarding its propensity score. That is, the weighting function is given by:

$$w^{NN}(i,j) = \begin{cases} 1 \text{ if } p_i - p_j = \min_j\{p_i - p_j\} \\ 0 \text{ otherwise} \end{cases} \tag{3.9}$$

with p_i and p_j as the propensity score for the i-th and j-th establishment, respectively.

It has also been suggested to use more than one non-treated establishment to construct the control group. Thereby establishments which are less similar are downweighted while higher weights are attached to those establishments which are more similar to the treated ones. The weighting function for this so called kernel matching is given by $w^{Kernel}(i,j) = K\left((p_i - p_j)/h\right)$ with $K(\cdot)$ as a kernel function, e.g. the Gaussian normal distribution, and h as a bandwidth parameter which can be calculated using Silverman's rule of thumb: $h = \frac{0.9 \min\{SD, IQR/1.34\}}{N^{1/5}}$.[19]

An alternative estimation strategy to this non-parametrical matching approach are simple regression models, e.g. in our case where the dependent variable is binary, a Probit model. Again we cite Angrist (1998) who notes that "the differences between regression and matching strategies for the estimation of treatment effects are partly cosmetic," since the approximation of any functional form imposed by a parametrical model like the Probit model as well as the conditioning on the propensity score in the matching model can be improved by including more interaction terms in the estimation. Hence, the "manner in which covariates are accommodated is not the most important difference between the two models" but instead lies in the weighting scheme used to take the average between treated and controls at different values of the covariates.

Let us briefly sketch how regression models are related to the conditional independence assumptions given in (3.7).[20] We assume the following Probit model where y_{1i} indicates the outcome of the i-th establishment. Let us further assume that we neglect sample selection on observables and instead, additionally to an intercept, only include a dummy variable D_i which indicates participation in a public subsidy:

$$Y_i = \begin{cases} 1 \text{ if } Y_i^* = \beta_0 + \Delta D_i + \epsilon_i \geq 0 \\ 0 \text{ otherwise.} \end{cases} \tag{3.10}$$

[19] SD stands for the standard deviation, IQR for the interquartile range of the propensity score and N for the number of observations. Details can be found in Silverman (1986).

[20] See also chapter two of this thesis.

As already mentioned, the independence assumption $Y \perp D$ will unlikely hold within a non-experimental environment. In reality it is more likely that there are observable covariates which are related to the outcome and the participation decision. These omitted variables will be reflected in ϵ so that any estimator for Δ will be plagued by the omitted variable bias. If, however, the conditional independence assumption holds, and we also include observable covariates X in (3.10), then $D \perp \epsilon | X$ and Δ can be estimated free of any bias as it is the case for matching models. The only difference between the two approaches lies in the weighting schemes used to take the average between treated and controls at different values of the covariates.

Angrist (1998) has shown that the weights used by the matching estimator are proportional to the probability of treatment at each value of the covariates, whereas in the linear regression model the weights are proportional to the variance of the treatment status conditional on the covariates. He has further shown that the weights used by matching and regression models will be roughly proportional if the values of $p(D_i = 1|X_i)$ are all less than 0.5. On the other hand, if the values of $p(D_i = 1|X_i)$ are all larger than 0.5, the weights are negatively correlated and hence the results of the matching and regression approach might differ.

3.3.2 Estimation Results

In the following we will present the results of a matching procedure and compare them with the results of a simple Probit model. Our treatment variable, D_i, is the fact whether the i-th establishment has received a public subsidy during 1997/1998 and our outcome variable, Y_i, the decision of establishments to improve or even to introduce new products during 1999/2000 (→ broad and narrow innovation concept). The first step of any matching procedure consists of an estimation of the propensity score $p(D_i = 1|X)$. To this aim we conducted a Logit estimation with the binomial treatment variable D_i as the dependent variable. It is important to note that this Logit model has no behavioral interpretation and thus needs no theoretical justification. The only objective is to balance the distribution of X between treated and non-treated establishments. Additionally to the variables contained in table 3.4, we included the squared number of employees, industrial and regional dummies.[21]

[21] We use regional dummies for every federal state. Due to an insufficient number of observations we only considered 7 industry sectors.

Since the Logit model has no theoretical underpinning, we will not present the detailed results of these estimations, but only some diagnostics to assess the quality of the estimated propensity score in the appendix (table B.4). In particular we will report LR-tests for the overall significance and LR-tests for industrial and regional dummies and explanatory variables. Along with the pseudo R^2 measure we also computed as an additional intuitive measure the so called hit-rate of each model compared to the naive hit-rate. The hit rate is defined as the percentage of correctly predicted observations. Thereby it is assumed that if $\widehat{p}(D_i = 1|X)$ exceeds the share of $D_i = 1$ observations in the dataset, $\widehat{D}_i = 1$. A naive proceeding would consist in simply using the share of $D_i = 1$ observations in the dataset as the probability to predict $\widehat{D}_i = 1$. The higher the hit rate compared to the naive hit rate the better the fit of the model.

Looking at the figures in table B.4 one can see that all estimated models are overall significant and that the pseudo R^2 ranges from 0.09 for East German to 0.24 for small and medium sized establishments. The hit rates as compared to a naive prediction of the dependent variable reveal a poor predictive power of the models. More important with regard to the validity of the subsequent matching, however, is the question whether the balancing property is fulfilled. A first impression is obtained by looking at kernel plots of the unbounded propensity score, i.e. the underlying latent variable of the Logit model. Ideally, we would like these kernels to look the same in the group of treated and non-treated establishments. Figure B.1 in the appendix, however, reveals that this is only the case for East German establishments. The impression of heterogeneity between treated and control establishments is confirmed if we look at the share of subsidized and non-subsidized establishments which had to be dropped due to the common support restriction. The figures in table 3.5 range from 8% in the case of treated micro establishments to 43% and 44% for non-treated and treated large establishments respectively.

For the estimation of the average treatment effect on the treated we used several approaches. We first conducted a one-to-one propensity score matching with common support restriction and replacement, i.e. each non-treated establishment may serve more than one time as a matched partner. In order to improve the matching quality, in a second step we additionally conditioned on the industrial sector.[22] We conducted the estimation for the whole sample and also separately

[22]Note that due to an insufficient number of observations conditioning on the industrial sector was not possible for different size groups.

for West and East Germany and for different size classes. Finally we repeated all estimations but now using a Gauss kernel matching.

Table 3.5: Some Matching Diagnostics

	Dropped Establishments due to Common Support (%)		Absolute Standardized Bias of Observable Covariates (%)		
	Treated	Controls	Before	After Matching[a]	
All Establishments	0.19	0.17	0.19	0.05	0.15
West Germany	0.24	0.15	0.28	0.12	0.11
East Germany	0.20	0.17	0.21	0.07	0.16
Micro Establishments	0.08	0.21	0.17	0.08	n.a.[b]
SME	0.18	0.31	0.18	0.07	n.a.[b]
Large Establishments	0.44	0.43	0.21	0.20	n.a.[b]

[a] One-to-one matching with replacement conditional on the propensity score and additionally on the industry sector (second column).
[b] Conditioning on the industrial sector not possible due to an insufficient number of observations.

Table 3.6 contains the results for the pooled estimation. Several points are worth mentioning. First, all estimated matching models yield a positive and significant average treatment effect on the treated ranging from 0.05 to 0.08 for the broad and from 0.11 to 0.12 for the narrow innovation concept. Hence, accounting for observable covariates, public subsidies increase the probability to improve or introduce new products by about 5 to 8% and if one focuses only on new products by even 11 to 12%. Thereby the results become more significant if we condition not only on the propensity score but also on industrial affiliation. Using more than one non-treated establishment as matched partners, as it is the case with the kernel matching, increases the effects and their significance. Tables B.2 and B.3 in the appendix contain the analogous figures for West and East Germany and for different size classes, respectively. The major stylized facts may be summarized as follows:

- In nearly all cases the ATE is positive and significant
- The ATE is higher for the narrower defined innovation concept
- The ATE is higher for East than for West German establishments
- Subsidies have the highest effect on small and medium sized establishments

- Kernel matching yields higher and more significant effects than one-to-one matching
- Matching not only conditional on the propensity score but also on the industry sector yields higher and more significant effects
- The only insignificant and even negative ATE is obtained for large establishments.

The reliability of all matching estimators, however, heavily depends on the balancing property of the propensity score. In order to check the quality of the matching, in table 3.5 we report the absolute standardized percentage bias of the covariates between treated and control establishments before and after the matching.[23] The figures in table 3.5 reveal that before matching, the two groups are most heterogeneous in West Germany (0.28) and more homogenous in the group of micro establishment with less than 10 employees (0.17). After the matching, the differences are reduced considerably but still remain high for West Germany (0.11) and for large establishments (0.20).

Table 3.6: Matching Estimator of the Average Treatment Effect of the Treated

	All Establishments					
	Broad Innovation Concept			Narrow Innovation Concept		
	ATE	SE	t-Value	ATE	SE	t-Value
One-to-One Matching	0.05	0.04	1.16	0.11	0.04	2.97
Conditional One-to-One	0.07	0.04	1.66	0.12	0.04	3.12
Kernel Matching	0.06	0.03	1.92	0.11	0.02	5.38
Conditional Kernel	0.08	0.03	2.70	0.12	0.03	4.29

We already noted that an alternative proceeding to the matching approach consists in estimating a Probit model with the innovation decision as the dependent and subsidies among other covariates as the exogenous variables. And we also already noted that the results of these two approaches will be similar if the

[23]The absolute standardized bias was proposed by Rosenbaum and Rubin (1985) and is defined for a single covariate x as follows: $|(\overline{x}(1)-\overline{x}(0))/(\sqrt{[var(x(1))+var(x(0))]/2}|$ where $\overline{x}(1)$ $(var(x(1)))$ is the mean (variance) of the covariate in the group of treated and $\overline{x}(0)$ $(var(x(0)))$ the mean (variance) in the group of non-treated establishments. The figures in table 3.5 were obtained by taking the average over the covariates.

values of $p(D_i = 1|X_i)$ are less than 0.5.[24] Table B.4 in the appendix contains in the last row the means and standard deviations of $\widehat{p}(D_i = 1|X_i)$ which are below 0.5 for every sub-sample. Hence, we expect the previous results to be confirmed by the estimation results of the following Probit model:

$$Y_i = \begin{cases} 1 \text{ if } Y_i^* = \beta' X_i + \Delta D_i + \epsilon_i \geq 0 \\ 0 \text{ otherwise,} \end{cases} \tag{3.11}$$

where again Y_i indicates the innovation decision of the i-th establishment and D_i whether it received a public subsidy.

The interested reader can find in the appendix a list of covariates contained in X and some reasons for their inclusion in chapter one. We will not present all estimation results but only focus on the impact of subsidies, i.e. on the parameter estimates for Δ, and some diagnostics for the models. Table B.5 in the appendix contains along with various LR-tests the estimated parameters of Δ. To quantify the impact of public subsidies on innovations, we first of all report the marginal effect of D for a reference establishment according to:

$$\frac{\partial Y_i}{\partial D_i} = \Phi(\widehat{\beta}' X_i + \widehat{\Delta}) - \Phi(\widehat{\beta}' X_i). \tag{3.12}$$

Additionally we also calculate the average treatment effect on the treated given by:

$$ATE^{Probit} = \frac{1}{N} \left\{ \sum_{i=1}^{N} \Phi(\widehat{\beta}' X_i + \widehat{\Delta}) - \Phi(\widehat{\beta}' X_i) \right\}, \tag{3.13}$$

i.e. the average difference between the probability that establishments would have introduced innovations if they had received a subsidy and the probability that establishments would have been innovative under no subsidy.[25] The standard error of the marginal effect as well as the average treatment effect were calculated using the delta method. Details can be found in the appendix.

At this point it is sufficient to notice that our a priori expectations are confirmed, i.e. that the Probit models yield by and large the same estimates for the

[24] These results, however, are only valid for the case of linear models and thus may only be used as an approximation when a probit model is estimated.

[25] We adopted the following reference establishment: All continuous variables are assumed to take on their mean values. Additionally we assume that the representative establishment does not possess a R&D-department, is a private limited company, operates in the transportion/telecommunication sector and is located in Baden-Wuerttemberg and accordingly in Thuringia for the sub-sample of East German establishments.

average treatment effect on the treated as the matching models. For East German establishments e.g. the ATE under the matching model for the narrow defined innovation concept amounted to 6 to 12% depending on the concrete matching model used. The Probit model on the other hand reports in this case a marginal effect of about 11% and an average treatment effect of 7%. Under the assumption that the sample selection mechanism is solely due to observable covariates, the results thus seem to be robust.

3.4 Sample Selection on Unobservable Covariates

In the previous section we have applied various econometric methods which can be used to estimate the effect of a treatment on an outcome variable in the presence of sample selection due to observable covariates. The methods differed regarding their parametrical assumptions but had one basic principle in common which could be summarized as follows: If there are differences between treated and non-treated subjects which are visible in the data, then try to compare only those subjects which appear to be comparable in the sense of their observable attributes. Applying these methods revealed a significant and considerable positive impact of subsidies on innovation activities.

Figure 3.2: Sample Selection Mechanisms

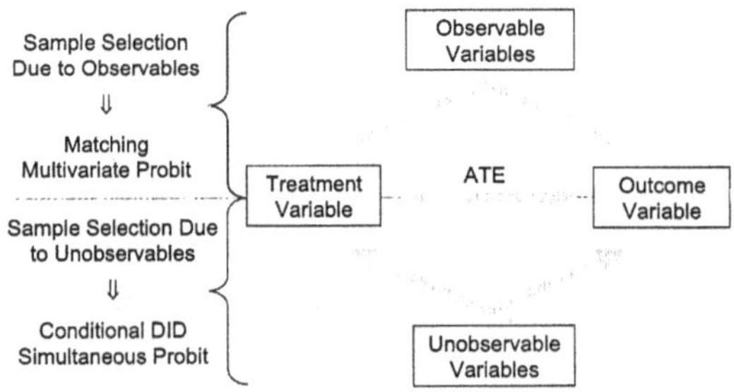

But what if, even after we have balanced the observable covariates, there are still differences between these two groups? Or stated differently, what if the conditional independence assumption which is the cornerstone of the matching as well as the Probit estimation does not hold and hence the sample selection mechanism is not only due to observable but also to unobservable covariates? In our application it could be the case e.g. that management ability or corporate culture play an important role in both determining innovation behavior and the decision whether an establishment will successfully apply for public subsidies or not. In this case the estimates would still be plagued by a "hidden" bias, upward we suspect, which could not be remedied by observable covariates, since both, management ability and corporate culture, are hardly quantifiable.[26]

In the following we will apply estimation strategies which can be used under these circumstances. Since in the last section we started with a nonparametrical approach, namely matching estimation, in this section it will be the other way round, i.e. we first discuss the simultaneous Probit model, which can also be interpreted as an instrumental variable approach, and then turn to its nonparametrical counterpart, the conditional difference-in-differences estimator (see also figure 3.2). The instrumental variable approach is attractive for at least one reason: The source of identifying information is transparent and accessible to testing as we will see.[27] This stands in contrast to both experimental and non-experimental approaches which, as Heckman, Lalonde, and Smith (1999) put it, "require assumptions that cannot be tested without collecting data specifically designed to test the assumptions of the model."

The use of Probit models was justified previously by the conditional independence assumption. Under this assumption, the included dummy variable for public subsidies, D_i, will not be correlated with the error term, ϵ_i, and hence the impact of subsidies on the outcome could be estimated free of any omitted variable bias. If the sample selection process, however, is due to unobservable characteristics, these unobservable variables will be reflected in ϵ_i and hence there will still be a potential correlation between D_i and ϵ_i. As a consequence, D_i will be endogenous. We will therefore start with an exogeneity test proposed by Rivers and Vuong (1988) and then try to find some appropriate instruments for this endogenous dummy variable.

The aim of Rivers and Vuong (1998) was actually to propose a simple two-step maximum likelihood procedure for estimating a recursive simultaneous equation model where one equation is of a Probit type and one of the continuous exogenous variables in fact determined by a second equation:

[26]We believe that sample selection due to unobservables plays a dominant role especially in the case of establishment data. The assumption of conditional independence assumption may be more plausible in the classical evaluation literature on active labor market programs where the researcher has access to exhaustive individual employee data.

[27]See Angrist (1991).

$$(3.14)\qquad \begin{aligned} y_{1i}^* &= x_{1i}'\beta_1 + \gamma y_{2i} + \epsilon_{1i} \\ y_{2i} &= x_{2i}'\beta_2 + \epsilon_{2i} \end{aligned}$$

with $y_{1i} = I(y_{1i}^* \geq 0)$ and $cov(\epsilon_{1i}, \epsilon_{2i}) = \sigma_{12}$.

Rivers and Vuong (1998) propose the following two-step estimation approach:

1. Regress y_{2i} on x_{2i} to obtain $\widehat{\beta}_2$ and the estimated residuals $\widehat{\epsilon}_{2i} = y_{2i} - x_{2i}'\widehat{\beta}_2$.

2. Conduct a Probit analysis with y_{1i} as the dependent and X_{1i}, y_{2i} and $\widehat{\epsilon}_{2i}$ as the explanatory variables.

The included residual in the second step, $\widehat{\epsilon}_{2i}$, accounts for the endogeneity of y_{2i} but additionally may also be used as an exogeneity test of y_{2i}. If namely the parameter of $\widehat{\epsilon}_{2i}$ is significant, the null hypothesis of exogeneity must be rejected. Thereby the significance of $\widehat{\epsilon}_{2i}$ may be tested by a simple t-test or alternatively by conducting a LR-test or a Wald test.[28] Another nice feature is the fact that under H_0 the distribution of ϵ_{2i} has no impact on the first Probit equation and hence the test is also valid without assuming normality of ϵ_{2i}.[29] It can therefore also be applied in cases where the second equation is of a Probit type as in our application:

$$(3.15)\qquad Y_i^* = \beta_1' X_{1i}' + \Delta D_i + \epsilon_{1i}$$

$$(3.16)\qquad D_i^* = \beta_2' X_{2i}' + \epsilon_{2i}$$

where the latent variables Y_i^* and D_i^* are connected with their observable counterparts via a threshold model, i.e. $Y_i = I(Y_i^* \geq 0)$ and $D_i = I(D_i^* \geq 0)$.

Equation (3.16) was estimated to obtain the propensity score for the matching while equation (3.15) was estimated as its nonparametrical counterpart. But now we allow these two equations to be connected with each other via the disturbances, ϵ_{1i} and ϵ_{2i}, for which we assume that

$$(3.17)\qquad E\begin{pmatrix} \epsilon_{1i} \\ \epsilon_{2i} \end{pmatrix} = \begin{pmatrix} 0 \\ 0 \end{pmatrix} \text{ and } Cov\begin{pmatrix} \epsilon_{1i} \\ \epsilon_{2i} \end{pmatrix} = \begin{pmatrix} 1 & \rho \\ \rho & 1 \end{pmatrix} = \Sigma.^{30}$$

The following table 3.7 contains the results of the Rivers and Vuong test of H_0 that the dummy variable D_i is exogenous. We conducted the test for different subsamples and also for the two different innovation concepts used for the outcome

[28]Details can be found in Rivers and Vuong (1988).

[29]See e.g. Wooldridge (2001).

[30]Note the necessary normalization of the two Probit equations.

equation. However, due to a small sample size we are not able to follow up large establishments. The same set of exogenous variables for X_{1i} and X_{2i} was used as before. One can see that for the narrow innovation concept all tests reject the hypothesis of exogeneity of the treatment decision.[31] For the broad innovation concept the LR-test always rejects exogeneity while the t-test and the Wald-test are only significant in the case of East German establishments. It thus seems that although we account for sample selection on observables, nevertheless the treatment decision is still endogenous and thus the two equations (3.15) and (3.16) need to be estimated simultaneously.

Table 3.7: Test for Exogeneity of the Treatment Decision

		Broad Innovation Concept			Narrow Innovation Concept		
	Test	Test-Value	df	p-Value	Test-Value	df	p-Value
All Establishments	t-Test	0.99	1,919	0.16	3.21	1,915	0.00
	LR-Test	59.06	1	0.00	60.52	1	0.00
	Wald-Test	0.98	1	0.32	10.31	1	0.00
West Germany	t-Test	0.33	1,106	0.37	1.79	1,104	0.04
	LR-Test	35.39	1	0.00	34.07	1	0.00
	Wald-Test	0.11	1	0.74	3.22	1	0.07
East Germany	t-Test	1.97	778	0.02	3.03	779	0.00
	LR-Test	28.98	1	0.00	29.33	1	0.00
	Wald-Test	3.88	1	0.05	9.16	1	0.00
Micro Establishments	t-Test	0.83	799	0.20	1.01	784	0.16
	LR-Test	13.59	1	0.00	15.01	1	0.00
	Wald-Test	0.68	1	0.41	1.03	1	0.31
SME	t-Test	1.09	793	0.14	3.09	789	0.00
	LR-Test	17.01	1	0.00	22.35	1	0.00
	Wald-Test	1.19	1	0.28	9.54	1	0.00

3.4.1 Simultaneous Probit Model

Equations (3.15) and (3.16) constitute a simultaneous equation model. Since both of the dependent variables are binomial, such models are also called mixed simultaneous equation models (MSEM) and since the observable counterpart of the latent endogenous dummy variable D_i^* enters equation (3.15), the model is of

[31] The only exception are micro establishments where at least the LR-test indicates endogeneity of public subsidies.

type II.[32] Type II MSEM exhibit not only the usual identification problem, which is ensured in our example if one looks at the list of variables included in X_{1i} and X_{2i}, but additionally also the so called coherency problem which makes some additional parameter restrictions necessary. The coherency problem arises since due to the inclusion of dummy endogenous variables as additional regressors, no explicit reduced form exists. Without going into detail how to derive these consistency restrictions, it is sufficient to notice that a two equation type II Probit model in order to be coherent must feature a recursive structure between the endogenous variables.[33] In our model this restriction is ensured by assuming that subsidies in 1997/1998 might have an impact on innovations in 1999/2000 but not vice versa.

The model in (3.15) and (3.16) will be estimated simultaneously by maximum likelihood although a first and tempting approach might be to use some seemingly "obvious" two-step approaches, where in a first step (3.16) is estimated by usual Probit and the obtained prediction $\Phi(\widehat{\beta}_2' x_{2i}')$ used as an instrument for D_i in a second step. However, such a proceeding is only valid if we included $E(D_i = 1)$ instead of D_i in equation (3.15).[34] Otherwise, since $\Phi(\cdot)$ is a nonlinear function, the expected value of D_i cannot be passed through and thus we have to derive the likelihood function of (3.15) and (3.16).

Let us to this aim additionally to (3.17) assume that:

$$\begin{pmatrix} \epsilon_{1i} \\ \epsilon_{2i} \end{pmatrix} \sim i.i.d.N(0, \Sigma), \tag{3.18}$$

i.e. our residuals follow a standard bivariate normal distribution which is solely determined by the correlation coefficient ρ. The likelihood function is then given by:[35]

$$L(\beta_1, \Delta, \beta_2, \rho) = \prod_{i=1}^{N} p_{11}^{Y_i D_i} p_{10}^{Y_i(1-D_i)} p_{01}^{(1-Y_i)D_i} p_{00}^{(1-Y_i)(1-D_i)}. \tag{3.19}$$

Let $\Phi_2(\cdot)$ be the bivariate cumulative standard normal distribution function.

[32] For a more thorough discussion of mixed simultaneous equation models see Blundell and Smith (1993, 1994). In part two of this thesis we will apply these models to test the skill biased technological change hypothesis.

[33] For details see e.g. Schmidt (1990) and also chapter six of this thesis.

[34] See e.g. Maddala (1983) or Wooldridge (2001). Angrist (1991), however, shows in a Monte Carlo experiment that especially in large samples, linear instrumental variable estimators perform nearly as well as the correctly specified ML-estimator.

[35] See e.g. Maddala (1983) but note the small misprint in the derivation of the likelihood function there!

Then:

$$
\begin{aligned}
(3.20) \qquad p_{11} &= p_{11}(Y_i = 1, D_i = 1) = \Phi_2(\beta_1' X_{1i}' + \Delta, \beta_2' X_{2i}', \rho) \\
p_{10} &= p_{10}(Y_i = 1, D_i = 0) = \Phi_2(\beta_1' X_{1i}', -\beta_2' X_{2i}', -\rho) \\
p_{01} &= p_{01}(Y_i = 0, D_i = 1) = \Phi_2(-\beta_1' X_{1i}' - \Delta, \beta_2' X_{2i}', -\rho) \\
p_{00} &= p_{00}(Y_i = 0, D_i = 0) = \Phi_2(-\beta_1' X_{1i}', -\beta_2' X_{2i}', \rho).
\end{aligned}
$$

Maximization of (3.19) yields finally the ML-estimates of the parameters.[36] Table B.6 in the appendix contains the appropriate estimation results. Recall once again that the Probit as well as the matching estimator yielded robust positive results for the treatment effect. However, the previously conducted exogeneity test pointed to the view that the treatment decision is still in fact endogenous and therefore the estimation results are plagued by an observational bias.

If we additionally account for sample selection due to unobservables, the results change dramatically. For the pooled estimation e.g. we found, using initially the matching approach, a significant average treatment effect on the treated of 5 to 8% for the broad and 11 to 12% for the narrow innovation concept. Now, this effect becomes negative and amounts to -20% and -17%, respectively! And the reason for this reversing sign is a positive and significant correlation between the treatment and innovation equation ($\rho = 0.51/0.64$). It thus seems that there are unobservable covariates which have a positive impact on both the subsidy and the innovation decision and which, if we do not account for them, will lead to an overestimation of the true treatment effect.

This result can be generalized to the other sub-samples as well. We found evidence that in all cases there is a substantial correlation between the two equations and hence the previous positive average treatment effects turn out to be overestimated. For West German and for small and medium sized establishments we now get negative results whereas in the case of micro establishments the results become insignificant. The only remarkable exception are East German establishments. Conditioning only on observable covariates as we did in the previous section yielded an average treatment effect on the treated of 2 to 11% for the broad and 6 to 12% for the narrow innovation concept. However, assuming that public subsidies are still endogenous, we now find a negative self selection due to unobservables (ρ = -0.61/-0.45) and hence the estimated effects even increase to 39% for the broad and 31% for the narrow innovation concept.

[36] The estimation was done using STATA.

An obvious question at this point concerns the reliability of these results. Having found convincing evidence for endogeneity of the treatment decision, the more crucial question is whether our simultaneous Probit model is able to mitigate this endogeneity. To this aim note that the model contained in equations (3.15) and (3.16) may be interpreted as an instrumental variable approach where the instruments used to model the endogenous dummy variable D_i are contained in X_{2i}. If one looks at the variables contained in X_{1i} and X_{2i} which can be found in the tables 3.4 and B.1, one can see that these are by and large the same exogenous variables. However, the variables contained in (3.15) refer to the year 1998 while the instruments refer to the year 1996. Our model is thus identified, but to be optimal and valid instruments, the exogenous variables in X_{2i} must satisfy the following two conditions: First, they must be significant in the participation equation (3.16) and second, they must be insignificant in the outcome equation (3.15).

In the following we conducted an informative test to check the validity of the instruments. In order to check whether X_{2i} is informative for D_i, we estimated a simple Probit with D_i as the dependent and X_{2i} as the independent variables and conducted a LR-test for the overall significance of X_{2i}. In a second step we tested whether X_{2i} has any explanatory power for the outcome equation by estimating a Probit with Y_i as the dependent and X_{1i} and X_{2i} as the independent variables. Again we conducted a LR-test for the joint significance of X_{2i}.[37]

Table 3.8 contains the results of these LR-tests and although this is just an intuitive test, we find convincing support for the validity of our instruments. Only for micro establishments using the broad innovation concept as the outcome variable, the instruments have also explanatory power in the outcome equation and thus may still be correlated with the error term. In all other cases, however, they are highly significant in the participation but insignificant in the outcome equation.

3.4.2 Nonparametrical Crosscheck

Our access to longitudinal data is limited. In particular we do not have regular information about the outcome decision. The foregoing time period for which such information is available refers to 1996/1997 and hence overlaps with the

[37]For a similar approach see Evans and Schwab (1995). This test is only an informal test because in the presence of endogenous regressors we would have to apply again a simultaneous estimator.

Table 3.8: Validity Test for the Instruments

		Broad Innovation Concept		Narrow Innovation Concept	
	Equation	χ^2-Value	p-Value	χ^2-Value	p-Value
All Establishments	Participation	31.64	0.00	31.64	0.00
	Outcome	4.34	0.63	7.07	0.31
West Germany	Participation	10.42	0.11	10.42	0.11
	Outcome	2.47	0.87	8.47	0.21
East Germany	Participation	29.42	0.00	29.42	0.00
	Outcome	5.44	0.49	8.59	0.20
Micro Establishments	Participation	14.64	0.02	14.64	0.02
	Outcome	11.63	0.07	6.95	0.33
SME	Participation	18.93	0.00	18.93	0.00
	Outcome	2.67	0.85	6.83	0.34

treatment period. The following considerations must therefore be treated with caution. One often used empirical strategy in the presence of individual specific but unobservable effects is to compare the innovation activity of an establishment after the treatment with the innovation activity of the same establishment before the treatment. The idea behind this strategy is that the outcome of an establishment before treatment might be a good proxy for the counterfactual outcome of no treatment.

Figure 3.3: The Before-After Approach

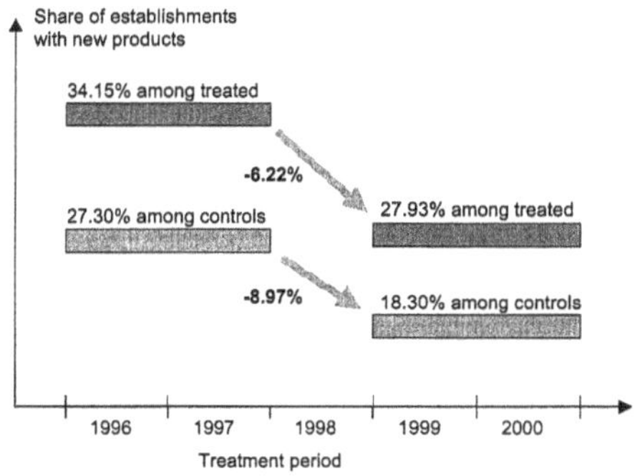

Figure 3.3 contains the shares of innovative establishments according to the narrow innovation concept before and after the treatment period. The simple before-after-estimator of the average treatment effect on the treated, Δ^{BA}, amounts to calculate the difference between the average outcome variable of treated establishments after treatment, $\overline{Y}_i^t(1)|D_i = 1$, and before treatment, $\overline{Y}_i^{t-1}(1)|D_i = 1$:

$$(3.21)\quad \Delta^{BA} = \overline{Y}_i^t(1)|D_i = 1 - \overline{Y}_i^{t-1}(1)|D_i = 1 = 27.93\% - 34.15\% = -6.22\%.$$

Applying this simple comparison yields thus a negative impact of subsidies on the innovation activities.

This simple before-after-estimator, however, has one major drawback.[38] It assumes that unobservable covariates average out over time, i.e. it assumes that the mean of the outcome variable is time invariant. This assumption is questionable especially in our context since innovation activities are sensitive to changing overall economic circumstances. An extension is the difference-in-differences estimator which also uses information about non-treated establishments and assumes that the following identifying assumption holds:

$$(3.22)\qquad E[Y_i^t(0) - Y_i^{t-1}(0)|D_i = 1] = E[Y_i^t(0) - Y_i^{t-1}(0)|D_i = 0].$$

Equation (3.22) states that the mean change in the outcome variable under no treatment is the same for subsidized and non-subsidized establishments.[39] If this assumption is valid, the difference-in-differences estimator can be applied:

$$(3.23)\qquad \begin{aligned} \Delta^{DID} &= \left([\overline{Y}_i^t(1)|D_i = 1] - [\overline{Y}_i^{t-1}(0)|D_i = 1]\right) \\ &- \left([\overline{Y}_i^t(0)|D_i = 0] - [\overline{Y}_i^{t-1}(0)|D_i = 0]\right) \end{aligned}$$

which in our case yields a positive impact, namely: -6.21% + 8.97% = 2.75%. The difference-in-differences estimator assumes that sample selection is solely due to unobservable, linear and time invariant effects. An extension is the conditional difference-in-differences estimator which additionally also controls for observables and ensures that only individuals are matched with the same observable characteristics.[40] In particular the conditional difference-in-differences estimator replaces $\left([\overline{Y}_i^t(0)|D_i = 0] - [\overline{Y}_i^{t-1}(0)|D_i = 0]\right)$ by its matched counterpart:

$$(3.24)\qquad \sum_{j\in\{D_j=0\}} w(i,j)\left([Y_i^t(0)|D_i = 0] - [Y_i^{t-1}(0)|D_i = 0]\right) N_0^{-1}$$

[38] See also the discussion in chapter two.

[39] For more details see Meyer (1995).

[40] See Heckman, Ichimura, Smith and Todd (1998) and for an application Hujer, Caliendo and Radić (2001).

with N_0 as the number of matched controls and $w(i,j)$ as the weight attached to the j-th non-participating establishment depending on the difference of the propensity scores for the i-th and j-th establishment. The following table 3.9 contains the results of the conditional difference-in-differences estimator. The estimation was conducted conditional on the same propensity score as previously and by using an one-to-one and Gaussian kernel matching approach, respectively.

Table 3.9: Conditional Difference-in-Differences Estimator (in %)

	Broad Innovation Concept		Narrow Innovation Concept	
	One-to-One	Kernel	One-to-One	Kernel
All Establishments	-2.27 (4.65)	0.87 (3.59)	4.44 (4.81)	4.56 (3.91)
West Germany	2.06 (8.80)	2.58 (5.83)	1.01 (11.54)	8.75 (6.74)
East Germany	-2.84 (5.99)	1.55 (3.90)	-3.78 (5.71)	2.94 (3.53)
Micro Establishments	-2.98 (7.65)	-1.77 (5.91)	-5.25 (7.30)	1.42 (4.74)
SME	9.84 (7.05)	4.98 (5.28)	9.33 (6.27)	8.58 (5.74)

Note: Standard errors in parentheses.

The results reveal that in most of the cases, except for small and medium sized establishments where we found a weak significant and positive impact, the average treatment effect on the treated is insignificant. We once again repeat that these results have to be treated with caution since due to data limitations the treatment period overlaps with the pre-treatment outcome period, but we also find support for our previous empirical findings, where the estimated treatment effects turned out to be insignificant or even negative if we additionally condition on unobservables.

3.5 Summary and Conclusions

The objective of this chapter was to add another piece of evidence to the ongoing political 'evergreen'-debate about the necessity to reduce public subsidies as one way to reduce public deficits. Although there is broad consensus about this point, the dispute which programs should be concerned is still ongoing and controversial. An often raised objection in that context is that public R&D subsidies should be excluded from this reduction since they are helpful to stimulate technological change and innovations. Despite the considerable amount spent in this area, the number of evaluations which try to assess the impact of such programs, however, is rather limited.

In this chapter we applied various microeconometric methods to overcome the inherent sample selection problem in estimating the effect of public subsidies on the innovation capacity of German establishments. We are tempted to accompany Klette, T. J. and Møen, J. and Griliches, Z. (2000), who stated that "in some ways we feel we are as confused as ever, but we believe we are confused on a higher level and about more important things." We started with a simple mean comparison which yielded significant positive impacts. We then accounted for sample selection due to observable characteristics by employing a parametrical multivariate Probit and a nonparametrical matching estimator. The result was a reduction in the effects which, however, still pointed to the view that public subsidies have a positive impact on innovations with differences for West and East Germany and different size classes.

Finally, we were able to show that especially with establishment data one has to take sample selection due to unobservables into account as well. Estimating a simultaneous Probit and a conditional difference-in-differences model changed the results dramatically. We only find positive effects for East Germany whereas in all other cases the results are at most insignificant.[41] Obviously there are various unobservable determinants which have a positive impact on the outcome as well as the participation decision. If we do not account for these effects, the impact of subsidies will be over-estimated. But what are the reasons for these negative microeconomic effects? One explanation might be that private capital is more efficient and productive than public subsidies. The better monitoring of private lenders could have an enhancing impact on the success probability of innovations. An alternative explanation might be that the relation between private and public R&D-spending is complementary, and that public subsidies crowd out private R&D-activities.[42] Whatever the true explanation might be, an upshot of all the estimations is the following: The more observable and unobservable factors one takes into account in order to make treated and control establishments more

[41] This finding is supported by the results of Czarnitzki (2001) and Almus and Czarnitzki (2002). The other findings which stand in contrast to Czarnitzki and Fier (2002) and to Licht and Stadler (2003) can only be explained by the different empirical strategy used which takes sample selection on unobservables into account as well.

[42] An alternative explanation that especially applies to smaller establishments is that the number of employees devoted to R&D is constraint and that there might be a trade-off between the effort devoted to R&D and the application for public subsidies.

comparable, the smaller the estimated treatment effect of subsidies on innovations becomes.

Of course we are aware that there are several limitations and drawbacks which have to be taken into account. We have only considered one kind of heterogeneity, namely with regard to West and East Germany and regarding different size classes. Other equally important types of heterogeneity which were not addressed in this study are, however, program heterogeneity and heterogeneity regarding the time structure. It could be the case that different measures have a different treatment effect and hence a multiple treatment framework would be more appropriate.[43] Another point concerns the assumed time structure. In this study we abstracted from the possibility that there might be more variable lag structures in the sense that subsidies granted in one year reveal their effect only with a delay in the subsequent years. Additionally it might also be the case that different "dosing schemes", i.e. the fact whether establishments received only once a subsidy or more than one time, might have a different impact on the outcome variable.

One has also to keep in mind that a microeconometric evaluation like this one is only a first step to evaluate the total net effect of public subsidies on the innovation activities. We have shown that only under the assumption of no spillover effects, these microeconometric findings can be generalized to the whole economy.

Despite all these caveats, we think that it is justified to close with the following quote of Lichtenberg (1984) who also found, after accounting for unobservable heterogeneity, unsatisfactory impacts of public R&D subsidies for the United States: "These findings thus make heavier the burden of proof on those who would claim that federal contract R&D makes a positive contribution to aggregate technical progress."

[43]See e.g. Lechner (2002b). Note, however, that limitations in the number of observations hindered us to conduct such an analysis.

Chapter 4

Companies: The Impact of Organizational Changes on Innovation Activities

4.1 Introduction

There is no doubt that innovations are a key factor for the future competitiveness and success of a single firm as well as for the economy as a whole. If one looks at the innovation capacity of German establishments, a disillusioning impression looms. The share of establishments e.g. which have introduced product innovations has decreased markedly over the last years. In this context the number of premonitory contributions rises which criticize the slackness of German establishments and ask for more flexibility and creativity.

One way out of this trap is the cutback of firm internal innovation restrains by implementing modern and flexible forms of workplace and human resource management (HRM) practices. These organizational changes can be characterized as a move away from the traditional 'Tayloristic' organization with its strong centralization of decision authority and its narrowly defined occupations. Instead, more and more establishments began to implement flexible workplace practices which are aimed at 'empowering' their employees by giving them more responsibility in a more flexible and less hierarchical environment. Such a 'Holistic' organization of work, however, needs to be complemented by flanking HRM practices.

There are a number of reasons indicating complementarities between different workplace and HRM practices. An obvious example is the necessity to offer employees training measures in connection with job rotation programmes. Implementing one single measure in isolation would lead to a suboptimal result as compared to the implementation of a coherent bundle of practices.

The aim of this chapter is to investigate the link between such flexible workplace and HRM practices on the one hand and the innovation propensity of establishments on the other, i.e. we seek to give an answer to the question whether more innovative establishments regarding their workplace and HRM practices are also more successful in implementing innovations. Thereby we will pay special attention to possible complementarities between different workplace and HRM practices. We will estimate a variety of econometric models by using a representative dataset for Germany, namely the IAB Establishment Panel Dataset conducted by the Institute for Employment Research ('IAB'). This chapter will therefore add one more piece of evidence to the ongoing debate about the impact of flexible workplace and HRM practices.

We proceed as follows: In the next section we briefly sketch the development of the innovation behavior in Germany during the last years. Section three is devoted to flexible workplace and HRM practices. We present some descriptive results about the implementation and diffusion of such practices and examine first links between them and the innovation behavior. Section four presents two econometric models, namely a trivariate Probit and a multinomial Logit model and discusses the estimation results while section five formulates some policy relevant conclusions.

4.2 Innovation Behavior in Germany

In the following we will present some first stylized facts about the innovation behavior of German establishments. Thereby we revert again to the IAB Establishment Panel conducted by the Institute for Employment Research ('IAB') at the German Federal Employment Office which represents a yearly survey of the demand side situation of the labor market.[1]

The IAB panel is organized in a modular form. There are topics covered annually like changes in the level and structure of employment and questions about the company's policy, e.g. information about the business volume and investment. Other topics are only covered irregularly, particularly information about innovations which have been at our disposal for the years 1993 and 1998.[2]

[1] For a more thorough discussion see Kölling (2000) or Bellmann (1997). See also chapter two of this thesis.

[2] Note that although questions regarding the innovation behavior of establishments were also posed

In this context the following questions were posed:

- Has your establishment improved an already offered product or service during the last two years? (*Improvement*)
- Has your establishment included a new product or service in its range of products during the last two years that already existed in the market? (*Incremental*)
- Has your establishment included a new product or service in its range of products during the last two years that were new for the market? (*Radical*)

Figure 4.1: Innovation Behavior of West German Establishments

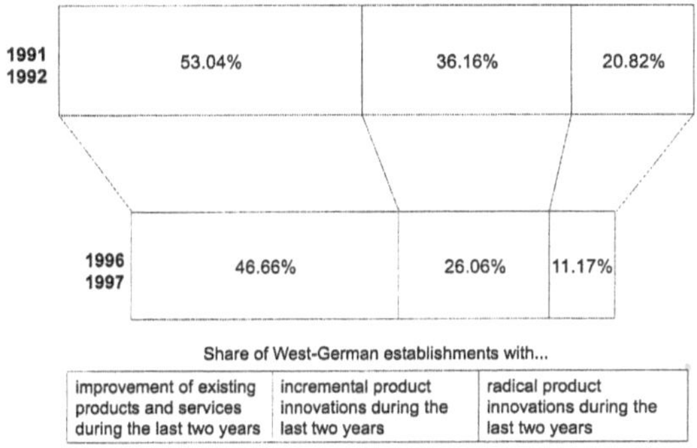

Looking at the share of establishments which have introduced the various forms of innovation reveals that the innovation activity of West German establishments has decreased considerably between 1991/1992 and 1996/1997 (see figure 4.1). The share of establishments which have introduced radical product innovations decreased from 20.82% to 11.17%. Improvements and incremental product innovations decreased by about the same magnitude.

This negative impression does not change if one differentiates between various size classes or different industry sectors. The figures in table 4.1 on page 91 disclose on the one hand that larger establishments are also the more innovative ones, e.g. the share of establishments with more than 500 employees which have either improved an existing product/service or introduced an incremental or even

in 2001, the IAB Establishment Panel contains no information about organizational changes for this year.

a radical product innovation amounts to 77.37% in 1991/1992, while the same figure for small and medium sized establishments (SME) with 5 to 500 employees amounts to 56.93%. On the other hand these figures also show that the innovation activity has dropped considerably from 1991/1992 to 1996/1997 across all size groups. Only 74.25% of all large establishments and 51.97% of all SME were still innovative in the years 1996/1997, a decrease of 3.12 and 5.96 percentage points, respectively.

Table 4.2 contains the same figures for different industry sectors: The industrial, service and industry related service sector. The first contains establishments in the primary, investment goods and consumer goods industry. The service sector is made up of establishments which belong to one of the following areas: Wholesale and retail, financing and insurance, transportation and telecommunication, legal advice, consulting, accounting, architecture and engineering. Finally, we excluded wholesale and retail establishments, i.e. the only distributive part of the service sector to obtain an industry related service sector. Although there are again considerable differences, with the industry as the most innovative sector, the general picture of decreasing innovations still remains.

East German establishments are only included since 1996 in the IAB panel, thus we are not able to depict their dynamic innovation behavior but only compare them with West German establishments for the years 1996/1997. Table 4.3 replicates the well known fact that East German establishments as compared to their West German counterparts are still in an inferior position in terms of innovation behavior.

This disillusioning empirical picture that we sketched for the time period until 1998 does not improve even in the following years. According to a study recently presented by the ifo Institute for Economic Research (see Penzkofer (2003)), the share of innovative establishments in the industry sector changed only slightly during the period from 1997 to 2000.[3] This share, however, dropped markedly after 2000 from 59% to 54% in 2001 and 53.0% in 2002. Studies conducted at the Centre for European Economic Research (see e.g. Janz, Ebling, Gottschalk, Peters, and Schmidt (2002)) arrive at a similar conclusion. They report that the share of innovative establishments in the industrial sector dropped from 67% in 1999 to 62% in 2000 and in the service sector from 63% to 60%. Recently the

[3] Innovative establishments are thereby defined as establishments which have either introduced a new product/service (product innovation) or conducted a process innovation, i.e. implemented a new and more efficient manufacturing method or production process.

Table 4.1: Share of West German Innovative Establishments for Different Size Classes

	1991/1992			**1996/1997**		
	≤ 5 Employees	5-500 Employees	> 500 Employees	≤ 5 Employees	5-500 Employees	> 500 Employees
Improvement	22.72%	46.82%	76.87%	19.60%	45.54%	74.38%
Incremental	19.55%	34.50%	45.82%	14.69%	26.34%	35.27%
Radical	9.39%	17.74%	31.21%	6.58%	9.98%	19.10%
Innovative at all	30.92%	56.93%	77.37%	14.91%	51.97%	74.25%

Table 4.2: Share of West German Innovative Establishments for Different Industry Sectors

	1991/1992			**1996/1997**		
	Industry Sector	Service Sector	Industry Related Services	Industry Sector	Service Sector	Industry Related Services
Improvement	76.16%	46.16%	61.09%	70.29%	40.97%	48.56%
Incremental	42.81%	42.93%	38.39%	33.52%	29.26%	28.69%
Radical	30.52%	19.53%	20.99%	17.52%	11.33%	10.88%
Innovative at all	80.38%	58.92%	62.23%	74.47%	49.63%	52.61%

Table 4.3: Share of Innovative West and East German Establishments in 1996/1997

	Germany	West Germany	East Germany
Improvement	39.64%	46.66%	33.43%
Incremental	24.74%	26.06%	23.57%
Radical	9.46%	11.17%	7.94%
Innovative at all	41.51%	46.21%	37.22%

European Commission (2002) presented a study which attested that German establishments only take a middle rated position among all European countries, the United States and Japan. Since there is no doubt that innovations are a key factor for the future success of a single establishment as well as for the economy as a whole, these premonitory signals have to be taken seriously. In order to foster the innovation process, however, first of all one has to understand the determinants of this process.

In this context a variety of theoretical and empirical contributions have emerged.[4] Traditional 'Schumpeterian' explanations emphasize the role of the size of establishments and the market structure and argue that especially larger firms with a high market power are capable of carrying out successful innovation activities.[5] Other studies in this context assume the existence of a 'knowledge function' where input factors like e.g. expenditures for R&D create patents and product innovations as an output.[6] Additional determinants that have often been considered in the literature are financial constraints (Harhoff (1998)), export activities (Ebling and Janz (1999)) and public R&D subsidies (Beise and Stahl (1998) or Czarnitzki and Fier (2002) (2002)).[7] Equally important factors are the human capital of establishments and the creativity of their employees.[8] To fully exploit the potential of high qualified employees, however, an appropriate workplace organization is needed. Thus more and more establishments recently introduced workplaces with

[4]See chapter two of this thesis for a more thorough discussion.

[5]For an empirical study on the impact of the market structure on product and process innovations in Germany see e.g. Flaig and Stadler (1994) or François, Favre, and Negassi (2002) for France.

[6]See e.g. Griliches (1990) who finds that R&D expenditure has a significant impact on innovation or Hall, Griliches, and Hausman (1986).

[7]For a review of the various determinants of the innovation process see e.g. Cohen (1995b).

[8]Lee, Florida, and Gates (2002) recently have analyzed which role the diversity of the local environment plays in the production of innovation. Their findings suggest that innovation at the regional level is positively associated with human capital and creativity.

a higher degree of functional flexibility and responsibility. In the following sections we will give an overview of the adoption of flexible workplace and HRM practices in Germany and try to assess their impact on the innovation process.

4.3 Workplace and HRM Practices

The eighties have experienced a boom in the field of popular management theory. Examples include the influencing work of Tom Peters and Robert Waterman (1982) "In search of Excellence" or Peter Drucker´s and Rosabeth Moss Kanter´s idea of 'Empowerment'. The upshot of these ideas was a call against overflowing tendencies of bureaucracy and centralization which more and more conflicted with the capacity of establishments to react quickly to changing economic circumstances and repressed their innovation capacity. Instead, companies were asked to decentralize decision authority and to 'empower' their employees by giving them more responsibility with lower restrictions and incentives to become more productive. All these changes in the workplace organization – more flexibility, more responsibility and less hierarchy – can be characterized as a move away from the traditional and centralistic 'Tayloristic' organization and have become known under the notion of 'holistic organization' or 'high-performance workplaces'.

There is a large theoretical literature on the determinants of such organizational changes. Lindbeck and Snower (1995, 2000) have developed a theoretical model with four major forces driving organizational changes: Increasing investments in more versatile capital, increasing investments in information and communication technology (ICT), more qualified work force and changing preferences in favor of versatile work. All these forces improve the trade-off between increasing rates of return from specialization in 'Tayloristic' organizations and rates of return from task complementarities in 'Holistic' organizations and thus also increase the probability to introduce more flexible workplace practices. We will undertake an empirical assessment of this model in the next chapter of this thesis.

With the introduction of such organizational changes, however, the insight began to grow that to fully exploit the benefits of changes in the organization of work, they have to be complemented by appropriate human resource management measures.[9] Only to implement workplaces with more responsibility and decision

[9]See e.g. Ichniowski, Shaw, and Prennushi (1997), Doeringer, Evans-Klock, and Terkla (1998) or Ichinowski, Kochan, Levine, Olson, and Strauss (1996). For a theoretical discussion of the concept of complementarity and supermodularity see e.g. Holmstrom and Milgrom (1994) or Milgrom and Roberts

authority e.g. and not complement these changes with a payment system which offers the individual employees incentives to exploit the offered scopes, may result in suboptimal results compared to the case where they were introduced in combination. Another persuasive example is the following: Employees which worry about loosing their jobs have only a limited interest in making productivity enhancing proposals which could rationalize their own jobs. If the relationship between the management and employees, however, is characterized by mutual trust, then again these changes may have the desired influence. In the following we will pay special attention to such complementarities between flexible workplace and HRM practices. Table 4.4 contains a description of the variables we have at hand to measure flexible workplace and HRM practices.[10]

There are a number of empirical studies which analyze complementarities between organizational changes and other input factors like e.g. investments in information and communication technology, capital and labor. They find empirical evidence for a significant leverage effect of organizational changes.[11] These papers mainly focus on the impact of organizational changes on the productivity of establishments or the labor demand.[12] The number of studies which try to quantify the impact of flexible workplace and HRM practices on the innovation capacity of establishments, on the other hand, is rather small. One exception is the paper by Michie and Sheehan (2000) who use the Workplace Industrial Relations Survey for the United Kingdom and were able to show that more flexible HRM practices are positively correlated with increasing investment in R&D. François, Favre, and Negassi (2002) conducted a similar analysis for France and demonstrated that the effects of organizational factors are even more important than the usually considered factors like size of the establishment and market structure. Recently Ernst (2003) presented a study where he analyzed the impact of organizational culture on the innovation performance of German establishments. He found that innovation-enhancing cultures have a positive impact on innovation success, whereas the effect of hierarchical cultures is negative.[13]

(1990).

[10]This classification closely corresponds to the one used by Ichniowski, Shaw, and Prennushi (1997) in their seminal paper on the effects of HRM practices on the productivity.

[11]See e.g. Hitt and Snir (1999) or Bresnahan, Brynjolfsson, and Hitt (2002).

[12]See e.g. for Germany Bauer (2001), Bertschek and Kaiser (2004) or Wolf and Zwick (2002) for the United States Black and Lynch (2001) or for Denmark Lundvall and Kristensen (1997). For the skill biased organizational change hypothesis see e.g. Falk (1999) or Hujer, Caliendo, and Radić (2002).

[13]It should be mentioned that there are a number of papers which analyze optimal organizations from a more business oriented perspective. Examples for Germany include Albers and Eggers (1991)

Table 4.4: Flexible Workplace and HRM Practices

Topic	Description of the Variable
Workplace Practices	
Flexibility	Decentralization of Responsibilities and Decision-Authority
Team Work	Introduction of Group-Workplaces with Own Responsibilities
Responsibility	Introduction of Units with Own Costing and Result Calculation
Job Rotating	Costs for Training and Education for Job Rotating Covered by Employer for Job Rotation
Human Resource Practices	
Hiring Practices	Plans for Manpower Requirements Set out in Writing
	Formal Plans for Staffing Existing
	Formal Job Description Existing
	Assessment of Job Performance Set out in Writing
Layoff Practices	Employees Laid off as a Reaction to Decreasing Sales in the Past
Job Security	Problems Expected by the Employer due to a too High Workforce
	Problems Expected by the Employer due to a High Labor Turnover
	Number of Layoffs relative to Total Number of Employees
Flexibility	Number of Part-Time Employees relative to Total Number of Employees
	Number of Workers on Temporary Loan Relative to Total Number of Employees
Mutual Trust	Workers Council Existing
Incentive Payment	Profit Participation Plans
	Wages above Collective Wage Agreement
Training	Company Finances Educational and Vocational Training
	Number of Employees Sent to Educational or Vocational Training Relative to Total Number of Employees

This chapter aims at adding one more piece of empirical evidence to the ongoing discussion about the impacts of flexible workplace and HRM practices on the innovation capacity of establishments. In the following we will first of all derive some stylized empirical facts about the diffusion of such practices among German establishments before presenting an empirical strategy how to take the complementarities between these practices into account and presenting some first simple descriptive statistics which point to a possible link to innovations. The interested reader can find several figures in the appendix which are intended to give him a first impression about the presence and diffusion of flexible workplace practices over time, for different size classes, different industry sectors and separated for

which analyze the so called "loose-tight principle" of organization.

West and East German establishments.[14] Instead of presenting them in the text, we will summarize the most important empirical patterns:

- The number of establishments which have introduced flexible workplace practices has declined over time.

- Most of these practices are introduced in the industry sector, followed by the sector of industry related services and the service sector.

- The size of the establishment is positively correlated with the propensity to introduce such practices.

- West German establishments are more likely to implement flexible workplace practices than their East German counterparts.

We already indicated that there are a couple of theoretical reasons to believe in the existence of complementarities between the introduction of flexible workplace and flanking HRM practices. Indeed, the major focus of this chapter lies in the assessment and quantification of such complementarities. However, the empirical implementation faces one major problem. The most obvious proceeding would consist in running a regression with an appropriate innovation indicator as the dependent variable and all dummy variables indicating the presence of flexible workplace and HRM practices as independent variables. In order to test the complementarity hypothesis explicitly, all possible interaction terms had to be included, too. In our case the number of exogenous variables therefore would amount to 19 dummy variables indicating the implementation of flexible workplace and HRM practices in isolation and 171 interaction terms to test if there are complementarities between two different practices, respectively. The high degree of multicollinearity between these practices makes such a proceeding, however, impossible.[15] As an alternative we seek to identify common clusters of establishments and conduct the econometric analysis with these cluster dummy variables as explanatory variables.[16] Each cluster should contain establishments which are

[14]Unfortunately most of the questions referring to HRM practices were only posed in 1998 so that we are not able to figure out the diffusion of these practices over time.

[15]A correlation matrix for the different flexible workplace and HRM practices is given in the appendix.

[16]For a similar proceeding see Ichniowski, Shaw, and Prennushi (1997) who identified clusters of establishments by "inspection" and additionally also applied a variety of statistical nominal scaling procedures. Wolf and Zwick (2002) conducted a factor analysis to overcome the multicollinearity problem.

similar regarding their workplace and HRM practices whereas the clusters themselves should be as heterogenous as possible. One standard method in this context is a cluster analysis. The interested reader can find the technical details of this cluster analysis in the appendix. At this point it is suffice to notice that applying such a cluster analysis yields two clusters of establishments.

In addition we also applied an alternative strategy to identify different establishment clusters by constructing an HRM-index. This HRM-index simply consists of the sum of implemented flexible workplace and HRM practices, i.e. establishments which have implemented more practices are assigned to a larger HRM-index and vice versa.[17] Different clusters of establishments were derived by partitioning them in four 25%-percentiles, respectively. This HRM-index is straightforward to interpret. For example, $HRM_1 = 1$ indicates that an establishment belongs to the lowest quartile of the least flexible establishments, whereas $HRM_4 = 1$ indicates the affiliation of the establishment to the top 25% of the most flexible establishments. Table 4.5 shows how these clusters constructed by using this HRM-index and the results of the cluster analysis differ from each other. One can clearly recognize that the HRM_4 cluster is the most innovative one regarding flexible workplace and HRM practices. For example, nearly 56% of all establishments belonging to this cluster decentralized their decision authority as compared to only 3.6% of all establishments in the lowest HRM_1 cluster. HRM_4 establishments are additionally more likely to introduce group-workplaces, offer their employees more responsibility and the possibility of job rotation. These establishments also feature a lower churning rate as indicated by the number of layoffs relative to the total number of employees and make more extensive use of temporary employment relations. Interestingly, the share of establishments expecting problems due to a too high labor workforce is higher among them, too. The differences in the shares between the two groups $Cluster_1$ and $Cluster_2$ are also pronounced.

The scattergram in figure 4.2 gives a graphical summarization of how the different clusters of establishments differ from each other. On the x-axis we depicted the sum of the implemented flexible HRM practices and on the y-axis the sum of the implemented flexible workplace practices. Besides a pronounced correla-

[17]It should be noted that some variables indicate the opposite of flexible HRM practices, e.g. if there are problems expected by the employer due to a high workforce or due to a high labor turnover or the fraction of layoffs relative to the total number of employees. These variables therefore enter the sum with a negative sign.

Table 4.5: Share of Establishments with Introduced Flexible Workplace and HRM Practices

	All	HRM_1	HRM_2	HRM_3	HRM_4	$Cluster_1$	$Cluster_2$
Workplace Practices							
Decentralization of responsibilities and decision-authority	23.65%	3.56%	10.54%	24.79%	55.69%	10.21%	42.08%
Introduction of group-workplaces with own responsibilities	16.55%	2.27%	6.14%	14.43%	43.34%	6.15%	30.8%
Introduction of units with own costing and result calculation	14.06%	2.07%	6.21%	13.46%	34.48%	5.93%	25.2%
Costs for training and education for job rotation covered by employer	9.53%	0.06%	2.59%	7.38%	28.07%	2.04%	19.79%
Human Resource Practices							
Plans for manpower requirements set out in writing	35.3%	2.39%	15.01%	47.18%	76.58%	7.52%	73.38%
Formal plans for staffing	36.91%	3.24%	18.24%	47.64%	78.53%	8.62%	75.72%
Formal job description	44.69%	6.54%	29.75%	61.55%	80.92%	18.85%	80.13%
Assessment of job performance set out in writing	34.28%	5.05%	16.82%	41.94%	73.29%	10.6%	66.74%
Employees laid off as a reaction to decreasing sales in the past	8.67%	14.24%	8.02%	6.54%	5.89%	9.26%	7.86%
Problems expected due to a too high workforce	21.42%	15.08%	19.15%	25.57%	25.87%	12.39%	33.79%
Problems expected due to a high labor turnover	2.62%	2.01%	2.72%	3.17%	2.59%	2.35%	2.99%
Number of layoffs relative to total number of employees in 1998	4.11%	13.41%	2.19%	1.08%	0.71%	6.6%	0.93%
Relative number of part-time employees in 1998	17.96%	9.81%	19.64%	21.71%	19.87%	20%	15.37%
Relative number of workers on temporary loan in 1998	0.59%	0.02%	0.158%	0.35%	1.77%	0.47%	0.75%
Workers council existing	50.35%	7.32%	36.02%	67.39%	85.02%	20.19%	88.7%
Profit participation plans	10.71%	0.97%	3.36%	9.13%	29.37%	3.5%	20.6%
Wages above collective wage agreement	33.84%	26.13%	30.78%	28.77%	43.75%	36.18%	32.01%
Company finances educational and vocational training	69.4%	17.9%	70.38%	90.99%	98.25%	49.78%	96.31%
Relative number of employees sent to vocational training in 1998	22.81%	2.65%	18.24%	26.94%	43.4%	16.79%	31.06%
Total number of establishments	6,182	1,545	1,546	1,545	1,546	3,575	2,607

Figure 4.2: Clusters of Establishments

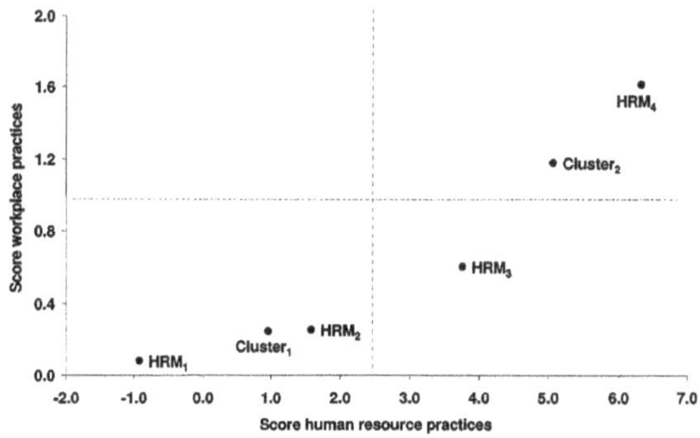

tion between both practices, one can again recognize that HRM_1, HRM_2 and $Cluster_1$ establishments are the least flexible ones regarding workplace and HRM practices while HRM_4 and $Cluster_2$ establishments are the most innovative ones. HRM_3 establishments are somewhere between these two polar cases.[18]

An interesting question at this point is whether establishments with flexible workplace and HRM practices differ in their innovation behavior. To this aim we divided the establishments in those which have either improved an already existing product/service or even introduced an incremental or radical product innovation and those which have not. The following figure 4.3 shows the position of these establishments in the workplace-HRM scattergram. This scattergram clearly points to the close relationship between flexible workplace, HRM practices and innovations. Those establishments with higher scores in both dimensions are also the more innovative ones.

Another piece of evidence for a positive link between introduced organizational changes and a higher innovation propensity can be obtained if one poses the following question: Is the fraction of innovative establishments larger among those

[18]Note that since some practices enter the score of HRM practices with a negative sign, the total sum can also become negative. At this point an alternative empirical strategy arises, namely the inclusion of the number of implemented flexible workplace and HRM practices as exogenous variables in a regression equation. In order to test the complementarity hypothesis, an interactive term defined as the product of these two indicators could be embedded, too. However, such an indicator is highly correlated with the sum of implemented workplace ($\rho = 0.83$) and HRM practices ($\rho = 0.56$). Due to this multicollinearity problem we abstained from this strategy.

Figure 4.3: Clusters of Innovative Establishments

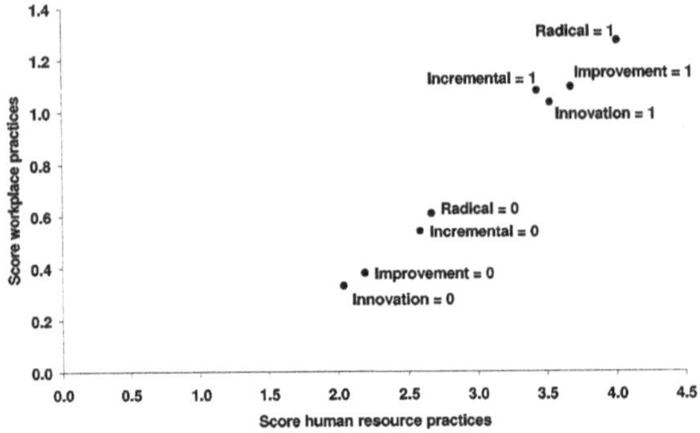

establishments which have implemented flexible workplace practices as compared to those establishments which have not? Figure 4.4 contains the answer to this question differentiated for different industry sectors. Looking at this figure one can first of all realize that the share of establishments which have introduced at least one of the flexible workplace practices listed in table 4.4 is largest in the industry sector, followed by the industry related service sector and finally the service sector. One can also clearly see that the share of innovative establishments is larger among those which have introduced flexible workplace practices compared to those which have not.

4.4 Econometric Modelling and Estimation Results

In the following we estimated two econometric models in order to test and quantify the effects of flexible workplace and HRM practices on the innovation propensity of establishments. Thereby we concentrate on establishments which participated in 1998 in the IAB Establishment Panel, i.e. due to fact that the information about the innovation behavior refer to the last two years, our dependent variable is the innovation behavior of West and East German establishments for the years 1996/1997. As mentioned previously, establishments were asked if they have improved an already existing product/service, introduced a product/service which was new for the own establishment but not new for the market (incremental product innovation) or if the innovation was a radical one, i.e. consisted of a prod-

Figure 4.4: Innovation and Flexible Workplace Practices

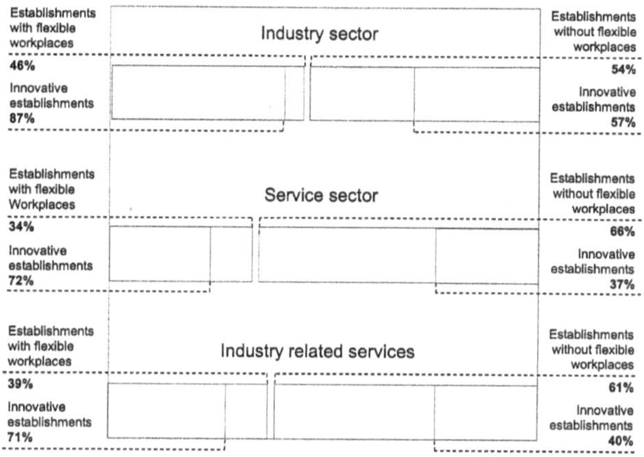

uct/service for which a new market had to be created. At first we will model these three decisions separately but allowing correlations between them by employing a trivariate Probit model. One advantage of this model as compared e.g. to an ordered Probit model is that it takes the heterogeneity of the parameters into account, i.e. it assumes that the effects of flexible workplace and HRM practices may be different for the different decisions at hand. One disadvantage, however, is the fact that we are not able to take overlapping decisions into account, e.g. if an establishment not only improves an already existing product/service but additionally introduces an incremental product innovation. To meet this feature of our dataset, in the second part of this section we will estimate a multinomial Logit model. For the estimation we excluded establishments belonging to the agricultural and public sector and estimated a variety of models differing in the included exogenous variables. We started with a basic model which only includes the dummy variables indicating flexible workplace and HRM practices, i.e. the *HRM* and the *Cluster* dummy variables, respectively. In an extension to this basic model we included industry sector, regional and size class dummies and finally also other exogenous variables which might have an impact on the innovation decision of establishments.[19] In the following sections, however, we will only

[19]The included variables are: Share of blue and white collar employees for qualified tasks, existence of an R&D and market research department, engagement in R&D cooperations with other institutions

report the estimation results of the *HRM* and *Cluster* dummy variables for the models which include the whole set of control variables.

4.4.1 Trivariate Probit Model

In this section we will estimate the three different innovation decisions the establishment faces simultaneously. To be more precise we will assume the following model:[20]

$$
\begin{aligned}
y_1 &= \begin{cases} 1 & \text{if } y_1^* = x_1'\beta_1 + \epsilon_1 > 0 \\ 0 & \text{otherwise} \end{cases} \\
y_2 &= \begin{cases} 1 & \text{if } y_2^* = x_2'\beta_2 + \epsilon_2 > 0 \\ 0 & \text{otherwise} \end{cases} \qquad (4.1) \\
y_3 &= \begin{cases} 1 & \text{if } y_3^* = x_3'\beta_3 + \epsilon_3 > 0 \\ 0 & \text{otherwise} \end{cases}
\end{aligned}
$$

where y_1 equals one if an establishment improves an already existing product/service and y_2 (y_3) equals one if the same establishment introduces an incremental (radical) product/service-innovation. Regarding the error terms we make the following assumptions:

$$
\epsilon = \begin{pmatrix} \epsilon_1 \\ \epsilon_2 \\ \epsilon_3 \end{pmatrix} \sim N(0, \Sigma), \qquad (4.2)
$$

i.e. we allow these three seemingly unrelated equations to be connected via possible correlations in the covariance matrix Σ. For reasons of identification we restrict the variances of the three Probit equations to be equal to one. The likelihood function of model (4.1) is given by:

$$
L = \prod p(Y_1 = y_1, Y_2 = y_2, Y_3 = y_3 | x_1, x_2, x_3, \Sigma) \qquad (4.3)
$$

where the probabilities that enter the likelihood function, e.g. $p(y_1 = 1, y_2 = 1, y_3 = 1 | x_1, x_2, x_3, \Sigma) = p(\epsilon_1 > -x_1'\beta_1, \epsilon_2 > -x_2'\beta_2, \epsilon_3 > -x_3'\beta_3) = p_{111}$, take on forms similar to the following:

$$
p_{111} = \int_{-x_1'\beta_1}^{+\infty} \int_{-x_2'\beta_2}^{+\infty} \int_{-x_3'\beta_3}^{+\infty} \phi_3(\epsilon_1, \epsilon_2, \epsilon_3 | x_1, x_2, x_3, \Sigma) d\epsilon_1 d\epsilon_2 d\epsilon_3. \qquad (4.4)
$$

like e.g. universities, indicators for the competition pressure, state of the technology and profitability, investments in ICT and dummy variables indicating different legal forms.

[20] We assume a representative establishment and therefore skip the i subindex for notational convenience.

The practical obstacle lies in the evaluation of the three-dimensional integral in (4.4) with $\phi_3(\cdot)$ as the trivariate normal density function. A solution to this problem is to use simulation methods to approximate this integral. One algorithm that has been found to be fast and accurate in this context is the GHK smooth recursive simulator which will be described in the following.[21]

Our aim is to evaluate expressions like the following:

$$p_{111} = p(\epsilon_1 > -x_1'\beta_1, \epsilon_2 > -x_2'\beta_2, \epsilon_3 > -x_3'\beta_3) \tag{4.5}$$

where the ϵ's are distributed as given in (4.2). The probability in (4.5) can be rewritten as a product of one unconditional and two conditional probabilities according to:

(4.6)

$$p(\epsilon_1 > -x_1'\beta_1)p(\epsilon_2 > -x_2'\beta_2|\epsilon_1 > -x_1'\beta_1)p(\epsilon_3 > -x_3'\beta_3|\epsilon_2 > -x_2'\beta_2, \epsilon_1 > -x_1'\beta_1).$$

Since Σ is a positive definite matrix using the Cholesky decomposition, a lower triangular matrix Λ can be found such that: $\Lambda\Lambda' = \Sigma$. Defining:

$$\nu = \begin{pmatrix} \nu_1 \\ \nu_2 \\ \nu_3 \end{pmatrix} \quad \text{with } \nu \sim N(0,1) \tag{4.7}$$

we can rewrite ϵ as $\epsilon = L\nu$ and arrive at the following relation between the ϵ's and the ν's:

$$\begin{aligned} \epsilon_1 &= \lambda_{11}\nu_1 \\ \epsilon_2 &= \lambda_{12}\nu_1 + \lambda_{22}\nu_2 \\ \epsilon_3 &= \lambda_{13}\nu_1 + \lambda_{23}\nu_2 + \lambda_{33}\nu_3 \end{aligned} \tag{4.8}$$

where λ_{ij} is the (i,j)-element in the Λ matrix. Using these relations, the product of unconditional and conditional probabilities in (4.6) can equivalently be written as:

(4.9)

$$\begin{aligned} p_{111} = \; & p\left(\nu_1 > \frac{-x_1'\beta_1}{\lambda_{11}}\right) \\ & p\left(\nu_2 > \frac{-x_2'\beta_2 - \lambda_{12}\nu_1}{\lambda_{22}} \middle| \nu_1 > \frac{-x_1'\beta_1}{\lambda_{11}}\right) \\ & p\left(\nu_3 > \frac{-x_3'\beta_3 - \lambda_{13}\nu_1 - \lambda_{23}\nu_2}{\lambda_{33}} \middle| \nu_1 > \frac{-x_1'\beta_1}{\lambda_{11}}, \nu_2 > \frac{-x_2'\beta_2 - \lambda_{12}\nu_1}{\lambda_{22}}\right). \end{aligned}$$

[21] More details can be found in Greene (2003) or McFadden and Ruud (1994). The estimation was done using STATA.

The advantage of (4.9) as compared to (4.6) is the fact that the ν's are independent normal distributed random variables and hence the probability p_{111} which we want to evaluate can be equivalently expressed as a product of independent but conditioned univariate cumulative density functions.

Assume now that ν_1^* and ν_2^* are realizations taken from truncated normal distributions with lower truncation points $\frac{-x_1'\beta_1}{\lambda_{11}}$ and $\frac{-x_2'\beta_2-\lambda_{12}\nu_1}{\lambda_{22}}$, respectively. Drawing samples from these truncated normal distribution ensures that the conditioning of the probabilities in (4.9) and therefore also in (4.6) is taken into account. Plugging ν_1^* and ν_2^* into (4.9), we can rewrite this expression as a product of only unconditional, univariate and independent probabilities:

(4.10)
$$\begin{aligned} p_{111} = & \; p\left(\nu_1 > \frac{-x_1'\beta_1}{\lambda_{11}}\right) \\ & \; p\left(\nu_2 > \frac{-x_2'\beta_2 - \lambda_{12}\nu_1^*}{\lambda_{22}}\right) \\ & \; p\left(\nu_3 > \frac{-x_3'\beta_3 - \lambda_{13}\nu_1^* - \lambda_{23}\nu_2^*}{\lambda_{33}}\right). \end{aligned}$$

The GHK simulator now generates a series of $d = 1, \ldots, D$ random observations of ν_1^{*d} and ν_2^{*d} so that the probability p_{111} that we seek for can be approximated by:

(4.11)
$$\hat{p}_{111} = \frac{1}{D}\sum_{i=1}^{N} \Phi\left(\frac{x_1'\beta_1}{\lambda_{11}}\right) \Phi\left(\frac{x_2'\beta_2 + \lambda_{12}\nu_1^{*d}}{\lambda_{22}}\right) \Phi\left(\frac{x_3'\beta_3 + \lambda_{13}\nu_1^{*d} + \lambda_{23}\nu_2^{*d}}{\lambda_{33}}\right).$$

All other probabilities can be calculated in an analogous way. Plugging them into the likelihood function in (4.3) standard maximization procedures can be applied to get estimates for the parameters.

In tables 4.6 and 4.7 we report the estimation results for the trivariate Probit model estimated using the GHK-simulator with $D = 25$ draws. As already noted, we only report the estimation results for the HRM-indices and the $Cluster$ dummy variable. Additionally, we report the results of LR-tests on the significance of the industry sector, region, size class dummies and a set of other exogenous variables. Notice that we included the same set of explanatory variables in every equation, i.e. $x_1 = x_2 = x_3$.

Table 4.6: **Estimation Results for the Trivariate Probit using the *HRM*-Index**

	Improvement	Incremental	Radical
HRM_2	0.158	0.227	0.509
	(0.060)	(0.064)	(0.097)
HRM_3	0.348	0.410	0.499
	(0.066)	(0.069)	(0.104)
HRM_4	0.470	0.528	0.690
	(0.074)	(0.075)	(0.108)
Correlations	ρ_{12}	ρ_{13}	ρ_{23}
	0.584	0.272	0.311
	(0.031)	(0.030)	(0.033)
LR-Overall	1,309 (0.000)	535.64 (0.000)	408.85 (0.000)
LR-Industry	143.96 (0.000)	85.21 (0.000)	41.74 (0.000)
LR-Region	22.92 (0.086)	25.76 (0.041)	23.60 (0.072)
LR-Size	33.48 (0.000)	2.10 (0.000)	0.26 (0.878)
LR-Controls	448.86 (0.000)	202.99 (0.000)	147.41 (0.000)

Note: Standard errors reported for the parameters and *p*-values for the *LR*-test.

In all estimated models the dummy variables indicating flexible workplace and HRM practices were found to be highly significant with a-priori expected signs. *LR*-tests reveal also a significant impact of the industry sector, region, size dummies and of the other explanatory variables. By including them sequentially in the models, the impact of the dummy variables of interest still remains significant but, however, the magnitude of the parameters decreases. Looking at the estimation results reveals that the parameter vectors for the different outcome variables differ substantially. The *HRM*-indices have the largest effect on the introduction of radical product innovations followed by incremental innovations and improvement of already existing products and services. Thereby the HRM_3-dummy has the largest impact followed by HRM_2 and HRM_1. For the *Cluster*-variable the differences between the parameters are not so pronounced. Another distinct feature of the estimated models are the positive and significant correlations between the different equations. Not taking this covariance structure into account and simply estimating separate Probit equations therefore would lead to inefficient parameter estimates.

Table 4.7: Estimation Results for the Trivariate Probit using the *Cluster*-Dummy

	Improvement	Incremental	Radical
$Cluster_2$	0.273	0.281	0.189
	(0.055)	(0.055)	(0.070)
Correlations	ρ_{12}	ρ_{13}	ρ_{23}
	0.582	0.265	0.314
	(0.031)	(0.029)	(0.032)
LR-Overall	853.52 (0.000)	358.44 (0.000)	329.25 (0.000)
LR-Industry	154.84 (0.000)	106.66 (0.000)	54.90 (0.000)
LR-Region	25.02 (0.050)	25.18 (0.050)	20.51 (0.153)
LR-Size	39.57 (0.000)	4.30 (0.117)	3.36 (0.187)
LR-Controls	449.15 (0.000)	219.25 (0.000)	162.09 (0.000)

Note: Standard errors reported for the parameters and p-values for the *LR*-test.

Once the estimated parameters are obtained the marginal effects of a change in an exogenous variable on the unconditional mean can be calculated:[22]

(4.12) $$E(y_j|x_i) = p(y_j = 1|x_i) = \Phi(x_i'\beta_j), \text{ for } j = 1, 2, 3.$$

Table 4.8 contains the marginal impacts of the *HRM* and *Cluster* dummies on the decision to introduce an incremental and a radical product innovation, respectively, given that all other exogenous variables take on their mean values. Introducing more flexible workplace practices and complementing them additionally by appropriate HRM practices considerable raises the probability to introduce product innovations. The fact e.g. that an establishment belongs to the upper 25%-percentile regarding flexible workplace and HRM practices increases its unconditional probability to introduce radical product innovations by about 11%!

Another interesting marginal effect can be derived if one looks at the conditional mean function, e.g. at the probability to introduce an incremental or radical product innovation, given that the establishment has already improved an existing product or service:

(4.13)
$$E(y_j|y_1 = 1, x) = p(y_i = 1|y_1 = 1, x) = \frac{\Phi_2(x_1'\beta_1, x_j'\beta_j, \rho_{1j})}{\Phi(x_j'\beta_j)}, \text{ for } j = 2 \text{ and } 3.$$

[22]See e.g. Greene (1996).

Table 4.8: Marginal Effect on Probabilities to Introduce Innovations

Dummy Variable	P(Dummy = 0)	P(Dummy = 1)	Marginal Effect
Unconditional Probability to Introduce Incremental Innovations			
HRM_2	0.221	0.294	0.073
HRM_3	0.208	0.343	0.135
HRM_4	0.199	0.376	0.177
$Cluster_2$	0.205	0.294	0.089
Unconditional Probability to Introduce Radical Innovations			
HRM_2	0.050	0.130	0.080
HRM_3	0.050	0.127	0.077
HRM_4	0.046	0.159	0.114
$Cluster_2$	0.060	0.086	0.026
Conditional Probability to Introduce Incremental Innovations			
HRM_2	0.735	0.755	0.020
HRM_3	0.725	0.784	0.059
HRM_4	0.719	0.801	0.082
$Cluster_2$	0.718	0.772	0.054
Conditional Probability to Introduce Radical Innovations			
HRM_2	0.629	0.655	0.026
HRM_3	0.610	0.699	0.089
HRM_4	0.602	0.720	0.118
$Cluster_2$	0.588	0.678	0.09

Although somewhat smaller than the impacts on the unconditional probability, the in that way calculated marginal effects again reveal considerable positive impacts of flexible workplace and HRM-practices on incremental and radical product innovations (see table 4.8). An interesting feature is that the absolute values of the conditional probabilities are by far larger for the conditional case than for the unconditional one, i.e. once establishments have already improved an existing product, the probability is quite high that they will in addition introduce an incremental or even radical product innovation.

4.4.2 Multinomial Logit Model

Estimating a trivariate Probit model in the previous section enabled us to take the parameter heterogeneity regarding the different innovation decisions like improvement, incremental and radical product innovations into account. One shortcoming, however, was the non-consideration of possible overlapping decisions, e.g. the case where an establishment not only improves but also introduces an incremental or radical product innovation. Table 4.9 shows that only 24.04% of all establishments, if any, choose one alternative in isolation. A large share of establishments (14.15%) improve existing products/services and additionally also introduce incremental product innovations. Around 5% of all establishments in the sample even carried out all three forms of innovation. These figures make clear that it is necessary to take overlapping decisions into account.

Table 4.9: Overlapping Innovation Decisions

Innovation Activity	Percentage	Dependent Variable
Do nothing	53.63%	$y = 0$
Only improvement	18.52%	$y = 1$
Only incremental product innovation	4.66%	$y = 2$
Only radical product innovation	0.86%	$y = 3$
Improvement and incremental product innovation	14.15%	$y = 4$
Improvement and radical product innovation	3.05%	$y = 3$
Incremental and radical product innovation	0.51%	$y = 3$
Improvement, incremental and radical product innovation	4.62%	$y = 5$

Establishments have the choice between three independent decisions: Improvement, incremental and radical product innovations. Combining them leads us to $2^3 = 8$ disjunctive alternatives. Since the relative frequencies for two alternatives, namely the introduction of only radical product innovations (0.86%) and incremental and radical product innovations (0.51%) are rather small, we decided to pool them with the decision to improve and introduce radical product innovations to one single alternative. Thus the choice set reduces from eight to six different alternatives including the base category $y = 0$.

Our explanatory variables are firm specific. The most simple model that can be applied in this context is the multinomial Logit model which assumes that the probability for an establishment to choose category $j = 0, 1, \ldots, 5$ is given by:[23]

$$p(y_i = j|x_i) = \frac{\exp(x_i'\beta_j)}{1 + \sum_{l=1}^{5} \exp(x_i'\beta_l)}, \text{ for } j = 0, 1, \ldots, 5. \tag{4.14}$$

Notice that for purposes of identification we have to restrict the β_0 vector to be equal to zero and hence treat $y = 0$ as the base category. Estimation of the pa-

[23]See e.g. Maddala (1983) or Franses and Paap (2001).

rameters can be done using maximum likelihood methods. Table 4.10 and table 4.11 contain the estimation results for two different models. The first includes the HRM-indices as well as the complete set of control variables while the second model uses the $Cluster$-dummy instead of the HRM-indices. Except for the HRM_2-dummy variable for $y = 1$ all variables are significant with a-priori expected signs.

Table 4.10: Estimation Results for the Multinomial Logit Model Using the HRM-Index

	$y=1$	$y=2$	$y=3$	$y=4$	$y=5$
HRM_2	0.157	0.463	1.280	0.408	0.718
	(0.131)	(0.184)	(0.299)	(0.152)	(0.296)
HRM_3	0.491	0.764	1.437	0.865	1.079
	(0.144)	(0.212)	(0.315)	(0.161)	(0.306)
HRM_4	0.659	1.198	1.878	1.123	1.736
	(0.161)	(0.241)	(0.326)	(0.178)	(0.313)
LR-Overall	643.42 (0.00)	158.82 (0.00)	387.30 (0.00)	661.45 (0.00)	497.75 (0.00)
LR-Industry	96.79 (0.000)	42.22 (0.000)	33.51 (0.004)	98.46 (0.000)	47.04 (0.000)
LR-Region	29.99 (0.012)	20.76 (0.145)	23.51 (0.074)	16.22 (0.368)	25.89 (0.039)
LR-Size	23.85 (0.000)	3.41 (0.182)	2.70 (0.259)	10.64 (0.005)	2.76 (0.251)
LR-Controls	211.13 (00)	43.57 (0.00)	157.07 (0.00)	282.19 (0.00)	232.26 (0.00)
Hausman-test	$Y \setminus \{1\}$	$Y \setminus \{2\}$	$Y \setminus \{3\}$	$Y \setminus \{4\}$	$Y \setminus \{5\}$
χ^2-value	0.00	-6.69	0.00	-0.00	0.00
p-value	1.00	1.00	1.00	1.00	1.00

Note: Standard errors reported for the parameters and p-values for the LR-tests.

The interpretation of the parameters is not straightforward since the marginal effects are highly nonlinear functions of the estimated parameters:

$$\frac{\partial p(y_i = j)}{\partial x_i} = p(y_i = j)\left(\beta_j - \sum_{l=0}^{J} p(y_i = l)\beta_l\right), \quad \text{for } j = 0, 1, \ldots, 5. \tag{4.15}$$

Notice again that $\beta_0 = 0$. Equation (4.15) indicates that the signs of the individual β_j parameters do not reflect the total net-effect of the exogenous variables on the probability $p(y_i = j)$. Indeed, even if β_{jk} e.g. is positive, the net-effect of x_{ik} on $p(y_i = j)$ might be negative! Instead of interpreting the parameters as marginal

Table 4.11: Estimation Results for the Multinomial Logit Model Using the *Cluster*-Dummy

	$y=1$	$y=2$	$y=3$	$y=4$	$y=5$
$Cluster_2$	0.629 (0.120)	0.767 (0.183)	0.670 (0.193)	0.687 (0.127)	0.925 (0.200)
LR-Overall	651.41 (0.00)	150.52 (0.00)	392.72 (0.00)	657.50 (0.00)	503.96 (0.00)
LR-Industry	100.34 (0.00)	50.77 (0.00)	41.47 (0.00)	113.22 (0.00)	52.96 (0.00)
LR-Region	31.62 (0.007)	19.95 (0.174)	23.15 (0.081)	17.24 (0.305)	25.68 (0.042)
LR-Size	22.81 (0.000)	3.45 (0.179)	4.12 (0.127)	13.67 (0.001)	4.01 (0.135)
LR-Controls	202.37 (0.00)	46.31 (0.00)	169.72 (0.00)	295.38 (0.00)	239.98 (0.00)
Hausman-test	$Y\setminus\{1\}$	$Y\setminus\{2\}$	$Y\setminus\{3\}$	$Y\setminus\{4\}$	$Y\setminus\{5\}$
χ^2-value	0.00	-4.21	0.00	0.00	0.00
p-value	1.00	1.00	1.00	1.00	1.00

Note: Standard errors reported for the parameters and p-values for the *LR*-tests.

effects one can consider the following log odds ratios:

$$\log \frac{p(y_i = j|x_i)}{p(y_i = 0|x_i)} = x_i'\beta_j. \tag{4.16}$$

Hence a positive β_{kj} parameter points out to a positive impact of x_{ik} on the probability to choose category j relative to the base category. In this sense all dummy variables indicating flexible workplace and HRM practices have a positive and significant impact on the probability that establishments become innovative at all compared to the alternative to do nothing.

Table 4.12 contains the marginal effects of the HRM and $Cluster_2$ dummy variables on the probabilities for the different alternatives according to (4.15) and given that all other exogenous variables take on their mean values. Again we find positive marginal impacts for the HRM and $Cluster$ dummy variables for all probabilities. The only exception is the impact of the HRM_1 variable on $y = 1$ which, however, did not turn out to be significant in the estimated model. By adding the appropriate probabilities one can calculate that the total marginal impact of the HRM_2, HRM_3 and HRM_4 dummies on the probability to introduce an incremental product innovation amounts to: 0.08, 0.086 and 0.131, respectively.[24] The corresponding total marginal impact of the $Cluster_2$ dummy

[24]The probability that an establishment introduces an incremental product innovation e.g. is given

on the probability to introduce an incremental product innovation is 0.037. The same exercise can be done for the marginal impacts of the HRM and $Cluster_2$ dummies for the probability to introduce radical product innovations. The results are: for $HRM_2 = 0.05$, for $HRM_3 = 0.18$, for $HRM_4 = 0.251$ and for $Cluster_2 = 0.11$. These numbers correspond by and large to the marginal impacts calculated for the trivariate Probit model.

Table 4.12: Marginal Effects for the Multinomial Logit Model

Dummy Variable	$\Delta p(y=1)$	$\Delta p(y=2)$	$\Delta p(y=3)$	$\Delta p(y=4)$	$\Delta p(y=5)$
HRM_2	-0.015	0.010	0.061	0.031	0.019
HRM_3	0.011	0.014	0.059	0.080	0.027
HRM_4	0.007	0.026	0.078	0.094	0.053
$Cluster_2$	0.058	0.017	0.015	0.056	0.022

One well known problem with multinomial Logit models is due to the fact that the log odds ratios for any two probabilities $p(y_i = j)$ and $p(y_i = k)$ are independent of the remaining alternatives. This property is known as the problem of independence from irrelevant alternatives (IIA). Intuitively, IIA means that the choice between two alternatives is not affected if not-chosen alternatives are removed from the choice set at hand. This assumption seems to be implausible in our context at first sight. The probability e.g. to introduce an incremental and a radical product innovation in combination may well be affected by the fact whether the establishment has the alternative to introduce a radical product innovation or not. In what follows we will therefore test if IIA holds for our sample and if we therefore can remain the estimation results.

IIA states that the probabilities for different alternatives do not depend on the choice set in which they are embedded, i.e. IIA assumes that relative probabilities of two alternatives remain the same whether a third alternative is included or not. One way to test this assumption therefore consists in omitting one or more alternatives, re-estimate the model and test if the parameter estimates differ substantially from each other or not. Tables 4.10 and 4.11 also contain the Hausman-tests for the different models where at each time one alternative was excluded from the choice set. For example, $Y \setminus \{1\}$ means that we have re-estimated the model after

by the sum of the following probabilities: $p(y=2) + p(y=3) + p(y=4) + p(y=5)$.

excluding category 1 out of the choice set. In all cases the differences between the restricted and unrestricted estimated parameter vector were not significantly different from zero. In some cases, e.g. where we excluded $y = 2$ from the choice set, the test statistic even is negative which is a sign that the IIA holds in our application.[25]

4.5 Summary and Conclusions

The study conducted in this chapter was motivated by the alarming slow-down of the innovation capacity of German establishments. In the second section we reported a decreasing share of establishments which have improved products/services in a magnitude of 6.4 percentage points between 1991/1992 and 1996/1997. In almost the same manner the share of establishments with incremental or radical product innovations decreased by 10.1 and 9.7 percentage points, respectively. An often raised resort out of this situation refers to the crucial role of human capital and the adoption of new models of work organization which entail a rising productivity and innovation propensity. Table 4.13 shows how the adoption of flexible workplace and HRM practices is connected to an improving innovation position. Although the most flexible establishments regarding workplace and HRM practices were not spared by a decreasing innovation propensity, the slow-down was somehow slower among them. Indeed the share of HRM_4 establishments with product improvements even increased during the considered time period by around 2%.

These simple descriptive results are also confirmed by a variety of econometric models which we estimated in the fourth section and which demonstrated that flexible workplace and HRM practices are able to raise the innovation probability considerably.[26] At this point a striking paradox arises.[27] There is a growing number of empirical studies, including this one, which document a positive impact of such practices on the productivity and also the innovation capacity. Therefore, in reality we would expect to observe a growing number of establishments which

[25] A negative sign of the Hausman test is indicates a lack of positive semidefiniteness of the covariance matrix. A solution would be to replace this matrix by an alternative statistic which always leads to a positive value. Hausman and McFadden (1984) note, however, that the corresponding value is unlikely to be so large as to come close to any reasonable critical value.

[26] Although the so obtained results are very robust regarding different econometric models and specifications, one drawback is due to a possible endogeneity bias. The appendix to this chapter contains some speculation about this issue.

[27] See also Pil and MacDuffie (1996).

Table 4.13: Share of West German Innovative Establishments

Dummy variable	Improvement 91-92	Improvement 96-97	Improvement Δ	Incremental 91-92	Incremental 96-97	Incremental Δ	Radical 91-92	Radical 96-97	Radical Δ
HRM_1	26.0 %	20.5%	-5.5%	25.0%	11.8%	-13.2%	12.5%	2.3%	-10.3%
HRM_2	36.3%	33.7%	-2.6%	27.3%	16.5%	-10.8%	12.7%	7.1%	-5.7%
HRM_3	55.1%	51.6%	-3.5%	36.6%	26.2%	-10.4%	17.1%	10.4%	-6.7%
HRM_4	71.5%	73.5%	2.1%	44.1%	37.9%	-6.2%	27.6%	19.8%	-7.8%
$Cluster_1$	32.7%	29.6%	-3.1%	27.3%	17.3%	-10.0%	13.0%	5.8%	-7.2%
$Cluster_2$	68.8%	68.4%	-0.4%	42.2%	33.1%	-9.1%	24.5%	16.7%	-7.8%

adopt such practices. In section three, however, we presented empirical evidence that the propensity to introduce such practices has decreased considerably over the last years. One explanation for this reluctant adoption of such practices lies in the costs incurred with such changes in the short run while the benefits are only gained in the long run. Such costs may be responsible that establishments do not implement optimal practices and remain in their existing suboptimal routines. Pil and MacDuffie (1996) refer to this phenomenon as a 'competence trap' and stress the importance of 'windows of opportunities' which make the implementation of various organizational changes more likely.

Figure 4.5: Benefits Associated with Flexible Workplace and HRM Practices

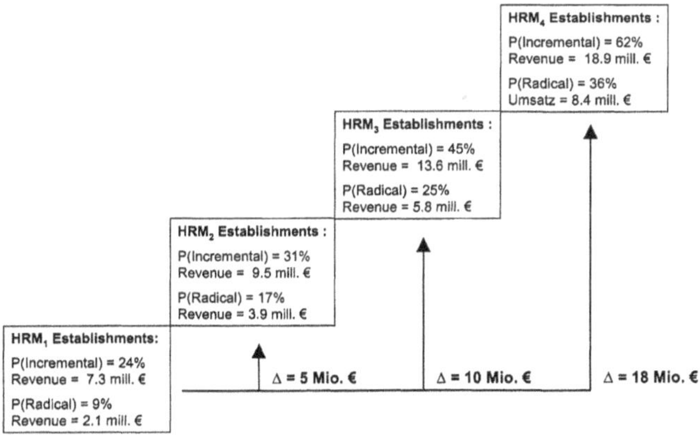

Referring to this cost argument, we want to conclude this chapter by presenting some calculations which monetize the benefits of flexible workplace and HRM practices. The establishments in our sample reported average revenues amounting to 277,613,000 €. The share of revenues induced by incremental and radical product innovations is on the average 10.91% and 3.37%, respectively. In the following we calculated the average probabilities for incremental and radical product innovations and the expected revenues generated by them for the different HRM-clusters of establishments. Looking at figure 4.5 one can see e.g. that the probability to introduce incremental (radical) product innovations for HRM_1-establishments, i.e. for the least flexible ones, amounts to around 24% (9%) with corresponding revenues of 7.3 (2.1) mill. €. In contrast, the most flexible establishments, i.e. those belonging to the top 25%-percentile, feature probabilities for introducing incremental and radical product innovations which amount to 62% and 36%, respectively. These increasing probabilities are also reflected in higher revenues of 18 mill. €. It is left to every establishment to decide whether these benefits outweigh the costs associated with the introduction of flexible workplace and HRM practices. We were, however, able to show that they exist.

Chapter 5

Inside the Black Box: Changes in the Organization of Work

5.1 Introduction

It is nowadays generally accepted that establishments with a higher degree of functional flexibility are in a better position to react and adapt promptly to new economic conditions and opportunities. This flexibility is especially necessary in an economic environment with increasing competition and pressure to adjust to changing circumstances.

In this context establishments have recently undertaken dramatic changes in their organizational structure characterized by a trend away from a centralized, so called 'Tayloristic' organization. Instead, more and more establishments introduced decentralized and flexible workplaces with more responsibilities for the individual worker and fewer hierarchical levels.

Before the introduction, establishments will weigh up the cost-benefit ratio arising from such organizational changes. Benefits include a reduction in a variety of cost categories, especially communication and information costs. Fewer hierarchical levels also imply a decrease in coordination and monitoring costs. And finally, since workplaces with more tasks and responsibilities offer employees more job satisfaction, guarantee more flexibility and additionally also might lead to complementarity effects, productivity gains can arise.[1]

[1]For a discussion about the productivity effects of organizational changes see e.g. Ichniowski, Shaw, and Prennushi (1997), Black and Lynch (2001) or for Germany Wolf and Zwick (2002). In the previous chapter we have documented that organizational changes increase the innovation propensity of establishments. Another point which will be discussed in the next chapter are the impacts of organizational

There are, however, costs which have to be considered and which are caused mainly by an increasing need for coordination among more autonomous units, a higher risk of duplication of information due to a lack of centralization, an increase of mistakes due to a lack of monitoring and decreasing returns to scale due to a lower specialization.

Various forces drive these changes. One obvious assumption is that the trade-off between costs and benefits of organizational changes improves with an increasing qualification structure of the employees. Higher qualified employees are more able to analyze the growing information flood caused by flatter hierarchies and to exploit the scope offered by workplaces with more responsibilities. Additionally, investment in more versatile capital and advances in information and communication technology will also spur organizational changes since these changes require flexible production methods and a better and instantaneous communication between the employees.

In what follows we will take a closer look at the process of organizational changes using the model presented in Lindbeck and Snower (2000) as a theoretical framework. Four driving forces were identified therein: Versatile capital, advances in information and communication technology, high qualified employees and changing worker preferences in favor of more varied jobs. The aim of this chapter is to assess the relative importance of these determinants empirically using an unique linked employer-employee dataset provided by the German Federal Employment Office.[2] This dataset contains detailed information on individual characteristics of employees and data on the establishments who employ them.

We proceed as follows: In the next section we will briefly sketch the Lindbeck/Snower model of organizational changes. Section three describes the dataset used for the estimation and presents some first descriptive results. Section four discusses the econometric modelling. We will estimate count data models with the total number of organizational changes as the dependent variable. Thereby special care is taken of the problem of excess zeros. Section five presents the estimation results while section six concludes.

---

changes on the labor demand.

[2]For a similar study see Carstensen (2002). Our approach, however, differs from this study in several aspects: First, our dataset is not restricted to the manufacturing sector in only one region of Germany, namely Lower Saxony. Second, we will explicitly model the organizational change decision whereas Carstensen (2002) only distinguishes between the two polar cases 'Tayloristic' and 'Holistic' and analyzes if these two clusters differ from each other regarding certain human resource management and workplace organization aspects.

5.2 The Lindbeck/Snower Model of Organizational Change

A fundamental dilemma concerning every organization is the trade-off between control and flexibility. Control is necessary to ensure that the organization keeps on efficiently pursuing its intended goals. Excessive control, however, suppresses creativity and flexibility which is necessary to adapt quickly to changing economic conditions and opportunities. Finding an appropriate balance between control and flexibility has therefore always been one of the major tasks of management. Changing times have witnessed different solutions to this problem.

The early twentieth century was characterized by mass production and exploitation of economies of scale through an excessive division of labor. Exemplary for this type of organization is the 'transfer line' technology introduced by Henry Ford. A central feature of this 'Tayloristic' organization is the single-purpose treatment of labor where employees are assigned to departments such as production, sales or accounting and perform within these departments highly specialized and narrowly defined tasks.[3]

Establishments nowadays seem to reverse this trend. They replace their single-purpose equipment by more flexible multitasking tools in order to be able to react more quickly and tailor-made to changing preferences of their customers. Related to these changes are new organizational strategies which have been introduced in the previous chapter and which have been summarized by Lindbeck and Snower (2000) in their seminal work under the notion of 'Holistic' organizations. Common and central features include the following:

- Emphasis on multitasking abilities, i.e. team work and job rotation
- Reduction in the number of management levels
- Continuous learning and development of complementary skills
- Decentralization of responsibility
- Direct participation of employees in decision making
- Emphasis on quality and customer orientation.

The common ground of these 'Holistic' organizational changes as compared to 'Tayloristic' organizations is the low degree of task specialization among employees and the utilization of complementarities between different tasks. Occupational

[3]See e.g. Snower (1998). Snower (1999) contains some anecdotic evidence.

barriers are breaking down and employees assigned to a broader range of activities. The rationale for these changes is that engagement in one activity can have a productivity enhancing effect on other activities. A sales person e.g. can gain valuable information about the customers which can be useful in the development of new products. Information gained in production may be useful in implementing accounting procedures and there are countless other examples indicating the existence of such complementarities.

The core problem of an establishment deciding about its optimal organization consists in balancing the trade-off between specialization and multitasking or, stated differently, in deciding how many 'Tayloristic' or 'Holistic' organizational elements to include in its own organization. In the following we will briefly summarize the major insights and features of the Lindbeck/Snower model which formalizes this decision problem. Let τ denote the degree of flexibility of a particular organization with $\tau \in [0,1]$. τ will be equal to one if an establishment has all characteristics of the polar case of a 'Holistic' organization mentioned previously and zero if it is organized in the traditional 'Tayloristic' way.

Increasing τ has a twofold effect on productivity y: First it increases the rates of return from inter-task complementarities and therefore y. On the other hand it also implies a decrease in the rates of return from specialization which would otherwise arise due to intra-task learning. As a consequence y will decrease. The optimal choice of an establishment depends on the functional form assumed for $y(\tau)$. Given $y_1(\tau)$ e.g. as depicted in figure 5.1 the optimal organization would be a 'Tayloristic' whereas for $y_3(\tau)$ it would be a 'Holistic' one. A production function $y_2(\tau)$ yields a mixed form with elements of both organizational forms.

Lindbeck and Snower (2000) argue that the shape of $y(\tau)$ and therefore the optimal choice of τ depends on a number of exogenous variables:

1. Changes in physical capital: Recent changes in capital investment have increased the versatility of machines across different tasks. As a consequence inter-task complementarities will arise.

2. Changes in information and communication technology: Increasing investment in ICT provide a rapid and cheap access to information which will facilitate and foster multitasking.

3. Changes in human capital: More qualified employees will be able to exploit the scopes offered by 'Holistic' organizations and to perform multiple tasks.

Figure 5.1: Productivity Effects of Organizational Changes

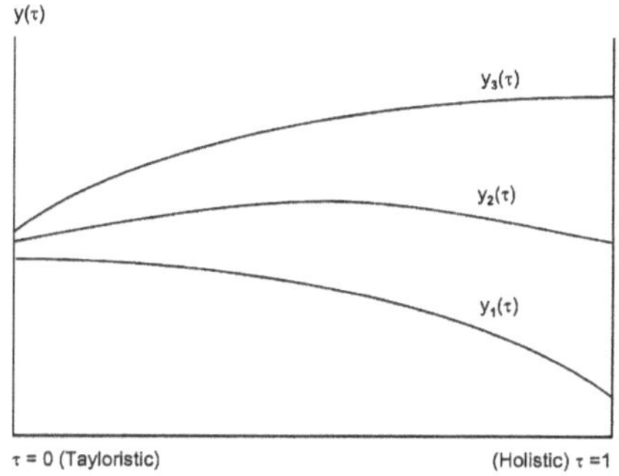

4. Changes in worker preferences: Many employees have a growing desire for more varying, challenging and stimulating jobs and therefore are more inclined to organizational changes.

Collecting these four exogenous variables in the vector $x = (x_1, x_2, x_3, x_4)$, the production function can be rewritten as:

$$y = y(\tau, x). \tag{5.1}$$

Due to the afore mentioned complementarities increases in x will also increase the benefits arising from organizational changes and thus shift the production function e.g. from $y_1(\tau)$ to $y_2(\tau)$ or even to $y_3(\tau)$.[4] We will capture this property by assuming that $y = y(\tau, x)$ is a continuous and supermodular function, i.e. increasing a component of x together with τ will increase y more than increasing x and τ separately.[5] With $\vee$ ($\wedge$) denoting the component wise maximum (minimum) of the two vectors (τ, x) and (τ', x'), the property of supermodularity can be formally stated as:

(5.2)

$$y\left((\tau, x) \vee (\tau', x')\right) - y\left(\tau', x'\right) \geq y\left(\tau, x\right) - y\left((\tau, x) \wedge (\tau', x')\right) \quad \forall (\tau, x), (\tau', x').$$

[4]Note that increases in x might even change the shape of $y(\tau)$ from concave to convex as demonstrated in Lindbeck and Snower (2000).

[5]Milgrom and Roberts (1990) or Holmstrom and Milgrom (1994) discuss the concept of supermodularity in more depth and give also an economic application. Note that in case of a continuous function supermodularity is equivalent to the condition that the cross-partial derivatives $\frac{\partial^2 y}{\partial \tau \partial x_i}$ are non-negative for all i, i.e. supermodularity reflects complementarity. Using the concept of supermodularity only facilitates the notation and highlights the main ideas of the Lindbeck/Snower model.

Due to the complementarities between τ and x in the production function the optimal choice of an establishment regarding its organizational form, τ^*, is a function of the internal Lindbeck/Snower factors. There are, however, factors which will only have an impact on the decision whether to adopt such practices or not but not on the productivity.[6] Examples for such factors might be e.g. competition pressure or an expected increase in business development which make organizational changes necessary and more profitable. We will capture this additional channel by collecting these factors in the vector z and assuming again that they exhibit a complementary relation with τ via a supermodular function $g(\tau, z)$.

The objective function of an establishment is then a function of the internal Lindbeck/Snower factors, x, and the external market factors, z, and can be written as the sum of the direct productivity gains due to organizational changes, $y(\tau, x)$, and the additional benefits from the introduction of such changes, $g(\tau, z)$:

$$\pi = f(\tau, x, z) = y(\tau, x) + g(\tau, z). \tag{5.3}$$

Note that the sum of two supermodular functions is again supermodular and thus $f(\cdot)$ supermodular in (τ, x) and (τ, z). Furthermore it can be shown that if $f(\cdot)$ is supermodular and $\tau^* = \arg\max\{f(\tau, x, z)|\tau \in \mathbb{R}\}$ a sublattice of the domain of f, the optimal choice of τ is a nondecreasing function of x and z, i.e. $\tau^* = \tau^*(x, z)$ with $\frac{\partial \tau^*}{\partial x} \geq 0$ and $\frac{\partial \tau^*}{\partial z} \geq 0$.[7] This result leads us to the following two hypotheses formulated in Lindbeck/Snower (2000) which will be tested empirically in the following:

Hypothesis 1 *Investment in more versatile capital and information and communication technology, rising shares of high skilled employees and changing worker preferences in favor of more varied jobs will increase the propensity and intensity of 'Holistic' organizational changes.*

Hypothesis 2 *External factors such as an increasing competition pressure, export share or an expected upswing in business development will increase the probability to introduce 'Holistic' organizational changes.*

5.3 Dataset and First Descriptive Results

For the empirical verification of the Lindbeck/Snower model we are able to use an unique linked employer-employee dataset provided by the German Federal Em-

[6] For a similar proceeding see Athey and Stern (1998).

[7] For a proof see e.g. Milgrom and Roberts (1990) or Holmstrom and Milgrom (1994).

ployment Office. Indeed, especially when analyzing the determinants of workplace organizations, it is necessary to take two sources of heterogeneity into account: Heterogeneity due to workers and heterogeneity due to firms. Linked employer-employee datasets provide the only way to cope with these issues. The availability of such datasets has been rather limited until recently when virtually an explosion in the use of such datasets initiated.[8]

For Germany a linked employer-employee dataset is available through the combination of the already introduced IAB Establishment Panel which contains detailed information about the employer side and the Employment Statistics Register covering the employees. The IAB Establishment Panel is conducted by the German Federal Employment Office.[9] Its population are all firms employing at least one employee subject to the compulsory social security system. The panel started out in 1993 with 4,365 West German establishments. Since 1996 East German establishments were also included in the dataset which contained about 15,526 establishments in 2000.

Employee information come from the Employment Statistics Register at the Federal Employment Office where all employers are obliged to report information about their staff which are subject to the compulsory social security system. Notifications include information about sex, age, qualification, occupation and wages of the employees. Employment groups such as public servants e.g. are not included in the dataset but still about 85% of all employees in Germany are covered.[10] Both datasets can be merged by an unique firm identification number.

Although termed 'panel', the IAB Establishment Panel is organized in a modular form where some issues are covered annually while others are not. Questions regarding organizational changes belong to this second category with the latest available information coming from the year 2000. Establishments were asked whether they have introduced one of the following organizational changes during the last two years, i.e. between 1998 and 1999:

- Introduction of group-workplaces with own responsibilities
- Reorganization of departments or functional areas
- Vocational training measures due to organizational changes

[8]See e.g. Abowd and Kramarz (1999a,1999b) Abowd and Kramarz (1999a) Abowd and Kramarz (1999b) or Haltiwanger et al. (1999).

[9]More details on the IAB Establishment Panel can be found in Bellmann (1997).

[10]See e.g. Bender, Haas, and Klose (2000) for detailed information about this dataset.

- Decentralization of responsibility and decision authority
- Reorganization of customer or supplier relations
- Introduction of units with own costing and result calculation
- Total quality management.

The items in this enumeration bear a good resemblance of the characteristics of a 'Holistic' organization as defined by Lindbeck/Snower: Team work, job rotation, reduction of management levels, decentralization of responsibility, continuous learning and greater emphasis on customer orientation. In the following we will use the total number of introduced organizational changes as the dependent variable. The number of implemented measures can be interpreted as a measure of the proximity of an establishment to the polar case of a 'Holistic' organization and thus as a proxy for the previously defined optimal degree of organizational flexibility τ^*.

Since the information regarding organizational changes refer to the years 1998 and 1999 we are using information from the 1998 wave of the IAB Establishment Panel to construct proxies for the Lindbeck/Snower factors (see also figure 5.2).[11] The first driving force of organizational changes mentioned by Lindbeck/Snower are changes in physical capital towards more flexible and versatile equipment which will be approximated by state of the technology used in 1998 ranked on a scale from 1 (up-to-date) to 5 (out-of-date). Changes in information and communication technology, as the second factor, are assumed to be indicated by investment in ICT (yes/no).

The next two factors identified by Lindbeck/Snower are changes in human capital and changes in the preferences of employees in favor of more varied, creative and challenging jobs. Both factors refer to employee characteristics and therefore are taken from the Employment Statistics Register. The former will be measured by the share of employees with an university or equivalent degree. The latter is more hardly to quantify. As a remedy we use the fraction of employees which are between 20 and 40 years old since it can be expected that especially younger

[11] Note that most of the information in the 1998 wave of the IAB Establishment Panel refer to the previous year 1997. In order to avoid an endogeneity bias we should have even moved another year backward. However, due to panel attrition our sample size would decrease considerably. The number of establishments which participated in 2000 and 1998 amounts to 8,104 but to only 6,294 establishments which participated in 2000 and 1997.

Figure 5.2: Time Schedule and Dependency Structure

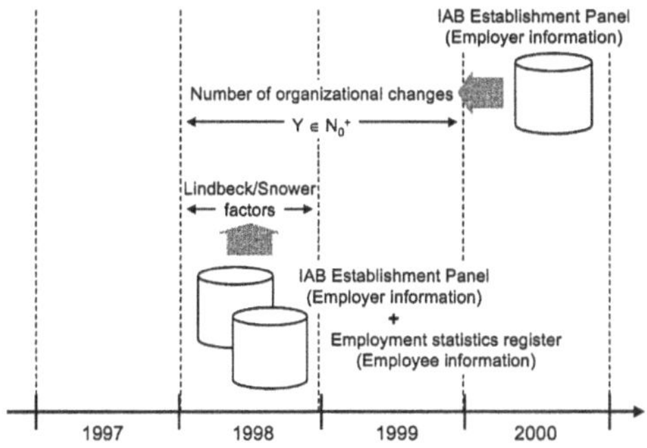

employees are not accustomed to a specific organization and therefore more willing to undergo organizational changes. Table D.1 in the appendix contains some summary statistics for these proxies. Merging the employer information for the years 2000 and 1998 with employee information for the year 1998 leaves us with 6,443 establishments. Excluding establishments from the agricultural and public sector finally reduces the sample size to 5,314 observations.

Before turning to the econometric modelling, in the following we will present some first descriptive results. Figure 5.3 contains the distribution of the number of organizational changes. One can see that most of the establishments have not implemented any measure at all. The mean of implemented organizational changes is equal to 1.41 with a standard deviation of 1.64. The figures presented in table D.1 in the appendix can be used to identify those establishment which have not implemented any organizational changes ($Y = 0$): Smaller establishments with less competition pressure, a smaller export share and a more unfavorable expected business development.

Figure 5.4 gives us a first impression about the impact of the Lindbeck/Snower factors on the number of organizational changes. We find only a modest positive impact of the state of technology while the connection between the other factors and the dependent variable is more pronounced. Among the establishments e.g. which have introduced all seven measures the share of establishments with investment in ICT is over 80% whereas establishments which have not introduced any organizational change are also those prone to not investing in ICT. The same

Figure 5.3: Number of Implemented Organizational Changes in 1998 or 1999

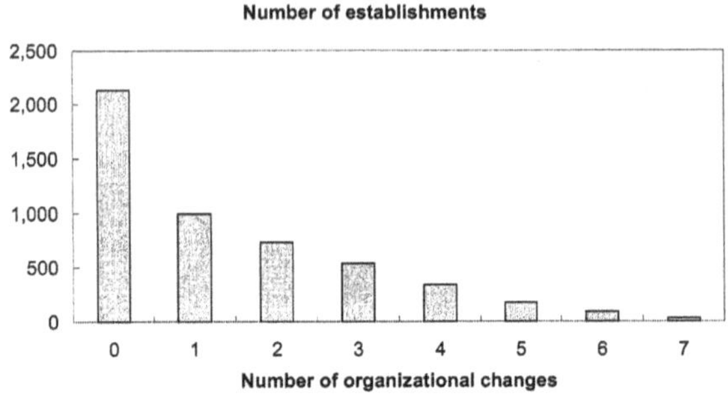

holds true for the share of high skilled and younger employees which are largest among those establishments with a higher number of organizational changes.

Figure 5.4: Lindbeck/Snower Factors of Organizational Changes

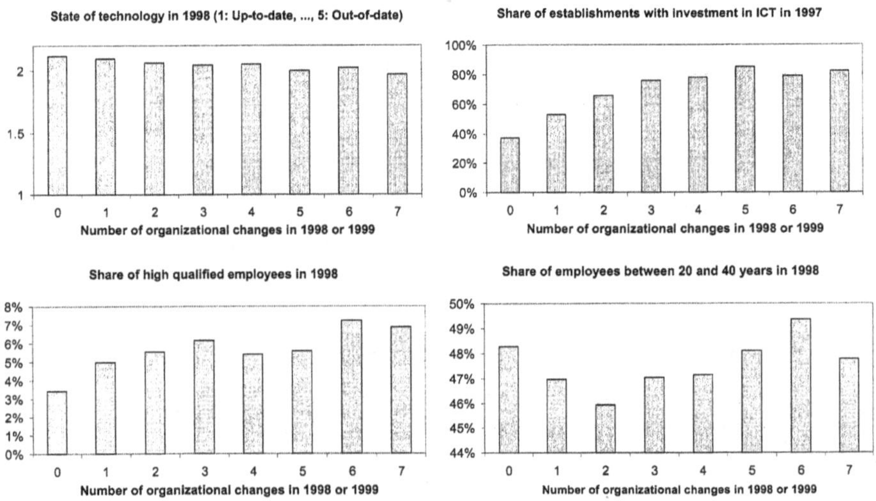

5.4 Econometric Modelling

The data on organizational changes presented in the previous chapter are typical count data which require the application of count data regression models.[12] Looking at figure 5.3, however, one might suspect that the distribution of the number of organizational changes is plagued by the presence of an extra portion of zeros which cannot be handled by standard count data models. One reason for such a prevalence might be that the zero outcomes of the dependent variable Y are not governed by one single data generating process (*dgp*) but that $Y = 0$ can be interpreted both as a qualitative as well as a quantitative information. In this case a model where $p(Y = 0)$ is restricted to follow the same *dgp* as the non-zero probabilities will be insufficient to explain the realizations of the dependent variable, especially of $Y = 0$.

In a standard Poisson model e.g. the probability that the dependent variable for the i-th individual, Y_i, takes on the value y_i is given by:

$$(5.4) \qquad p(Y_i = y_i|\lambda_i) = f_p(y_i) = \frac{\exp(-\lambda_i)\lambda_i^{y_i}}{y_i!} \text{ with } \lambda_i > 0 \text{ and } y_i \in N_0^+.$$

λ_i is the intensity parameter of the Poisson process with $\lambda_i = E(Y_i) = Var(Y_i)$. The last equality is referred to as equidispersion and will be discussed in the following in more depth. Fitting this simple model without covariates, for illustrative purposes only, leads to a clear under-representation of zero outcomes as can be seen from figure 5.5. Three extensions have been proposed in this context: With zeros, zero inflated and hurdle models.[13] The idea of with zeros models is to add an extra proportion of zeros, ω, to the parent distribution and decrease the remaining probabilities correspondingly to insure that all probabilities sum up to one. In case of a Poisson model the parent distribution is $f_p(y_i)$ thus leading to the following with zeros Poisson model:

$$(5.5) \qquad p(Y_i = y_i|\lambda_i, \omega) = \begin{cases} \omega + (1-\omega)f_p(0) \text{ for } y_i = 0 \\ (1-\omega)f_p(y_i) \text{ for } y_i \in N^+. \end{cases}$$

The predicted outcomes from this model, again only for illustrative purposes estimated without covariates, are also depicted in figure 5.5 and reveal a much better fit to the data. This visual impression is supported by a score test for zero

[12] See e.g. Cameron and Trivedi (1998,1999) or Winkelmann (1997) for an overview of count data regression models.

[13] See e.g. Mullahy (1986) or Crepon and Duguet (1997).

inflation in Poisson distributions proposed by van den Broek (1995). Under the null hypothesis of $\omega = 0$ and no covariates the test statistic

$$(5.6) \qquad S(\hat{\lambda}) = \frac{(n_0 - n\widehat{p_0})^2}{n\widehat{p_0}(1-\widehat{p_0}) - n\overline{y}\widehat{p_0}^2}$$

follows a chi-square distribution with 1 degree of freedom. Thereby n_0 and n denote the number of zero and total observations, $\widehat{p_0}$ is the estimated probability of $Y = 0$ which equals in a Poisson model without covariates $\exp(-\hat{\lambda})$ and $\overline{y}$ is the mean of the outcome variable. Computing $S(\hat{\lambda})$ using our dataset yields a value of 1,602.64 which clearly indicates the presence of excess zeros.

Figure 5.5: Observed and Predicted Counts of Organizational Changes

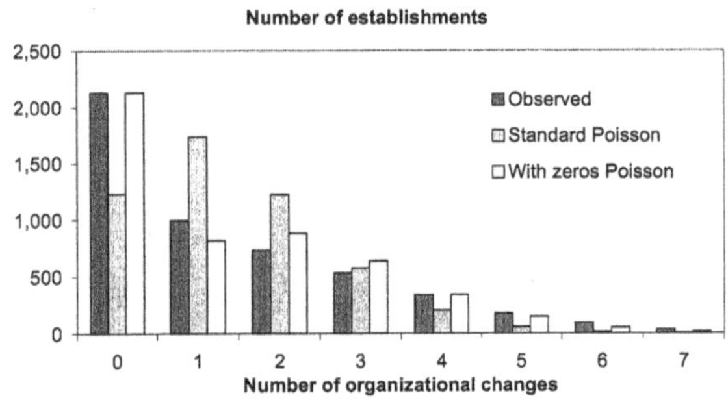

Covariates are typically included by assuming that the intensity parameter of the Poisson distribution is a function of the explanatory variables, i.e. $\lambda_i = E(Y_i) = Var(Y_i) = \exp(X_i\beta)$ where the last equality ensures that $\lambda_i > 0$. In with zeros models ω is restricted to be a constant. In contrast, zero inflated models are characterized by not only parameterizing λ_i but ω as well. Lambert (1992) proposed to use a standard Poisson distribution for Y and a logistic function for ω which leads to the zero inflated Poisson model:

$$(5.7) \qquad \begin{aligned} f_p(y_i) &= \frac{\exp(-\exp(X_i\beta))\exp(X_i\beta)^{y_i}}{y_i!} \\ \omega &= \frac{\exp(Z_i\gamma)}{1+\exp(Z_i\gamma)}. \end{aligned}$$

Thereby the covariates that affect the Poisson mean, X_i, may or may not be the same that affect the additional zero probability, i.e. Z_i.

Lambert (1992) gives an interpretation of the zero inflated model as a model with two 'states of the world' from which zeros can arise. In one state the outcome variable is always zero. In the second state zeros are rare and governed by the parent distribution. The decision which state is at work is thereby determined by a binary random process. In her application she analyzes a manufacturing process that is either under control so that the number of items with no defects is extremely rare and an imperfect state in which defects are more likely. Van den Broek (1995) gives another example: He analyzes the number of times HIV-infected men had an urinary tract infection. He argues that one proportion of patients has the physiological preconditions to develop such an infection while the other has not.

This interpretation, however, is questionable in our application. It seems unrealistic to assume that there are establishments which will *never* introduce organizational changes per se and it is hardly imaginable to find some characteristics for such 'change-resistent' establishments.[14] More reasonable is the assumption that establishments will first of all decide whether to introduce organizational changes or not. Given a positive first-stage decision, in a second step they decide on how many measures to introduce. The factors determining both decisions are thereby not restricted to be the same.

Hurdle models provide a way to model such a two-step decision process. In a first step a binomial random process following the distribution $f_1(y_i|\theta_1)$ determines whether the outcome variable is zero or not. If the realization is non-zero, i.e. the 'hurdle' has been crossed, a truncated-at-zero distribution, $\frac{f_2(y_i|\theta_2)}{1-f_2(0|\theta_2)}$, describes the non-zero outcomes. Thus:

$$p(Y_i = y_i) = \begin{cases} f_1(0|\theta_1) \text{ for } y_i = 0 \\ (1 - f_1(0|\theta_1)) \frac{f_2(y_i|\theta_2)}{1-f_2(0|\theta_2)} \text{ for } y_i \in N^+. \end{cases} \tag{5.8}$$

Multiplying the truncated distribution with $(1 - f_1(0|\theta_1))$ ensures that the probabilities sum up to one. θ_1 and θ_2 are thereby the parameters of the two distributions. The difference between zero inflated and hurdle models is that for the former the parent distribution is valid for the full range of $Y \in N_0^+$ and not just for strictly positive values as in the latter case.

The parameters $\begin{pmatrix} \theta_1 & \theta_2 \end{pmatrix}$ of a hurdle model are estimated by the method of

[14] An exception might be public authorities which have been excluded from the analysis, however.

maximum likelihood with the following general log-likelihood function:

(5.9)

$$\begin{aligned}\ln L &= \ln\left(\prod_{y_i=0} f_1(0|\theta_1) \prod_{y_i>0} (1-f_1(0|\theta_1)) \frac{f_2(Y_i=y_i|\theta_2)}{1-f_2(0|\theta_2)}\right) \\ &= \ln\left(\prod_{y_i=0} f_1(0|\theta_1) \prod_{y_i>0}(1-f_1(0|\theta_1)) \prod_{y_i>0} \frac{f_2(Y_i=y_i|\theta_2)}{1-f_2(0|\theta_2)}\right) \\ &= \left(\sum_{y_i=0} \ln f_1(0|\theta_1) + \sum_{y_i>0} \ln(1-f_1(0|\theta_1))\right) + \left(\sum_{y_i>0} \ln\left(\frac{f_2(Y_i=y_i|\theta_2)}{1-f_2(0|\theta_2)}\right)\right).\end{aligned}$$

The last equality makes clear that the parameters of a hurdle count data model can be obtained by a separate estimation of a binary random process and a truncated distribution for the positive observations. This follows from the fact that the two mechanisms $f_1(y_i|\theta_1)$ and $f_2(y_i|\theta_2)$ are assumed to be independent from each other.

If $f_1(y_i|\theta_1) = f_2(y_i|\theta_2)$ then the hurdle model collapses with the standard count data model. In this case the standard formulation yields a consistent and efficient estimator whereas the hurdle specification is only consistent. One possible way to test the hurdle against the basic count data specification is therefore a simple Hausman test.

Winkelmann (2003) gives an overview of possible combinations of functions for $f_1(y_i|\theta_1)$ and $f_2(y_i|\theta_2)$. A commonly used one is that both $f_1(y_i|\theta_1)$ and $f_2(y_i|\theta_2)$ follow a Poisson distribution but, however, with differing intensity parameters $\theta_1 = \lambda_1 = \exp(X_{1i}\beta_1)$ and $\theta_2 = \lambda_2 = \exp(X_{2i}\beta_2)$:

(5.10)

$$\begin{aligned}f_1(0|\beta_1) &= \exp(-\exp(X_{1i}\beta_1)) \\ f_2(Y_i=y_i|\beta_2) &= \frac{\exp(-\exp(X_{2i}\beta_2))\exp(X_{2i}\beta_2)^{y_i}}{y_i!}.\end{aligned}$$

Note at this point that the presence of excess zeros typically implies overdispersion with the variance of Y increasing with the mean. Not taking this overdispersion into account has similar consequences as the failure of the homoscedasticity assumption in the linear regression model: As long as the mean is correctly specified, the parameters are still consistently albeit not efficiently estimated.[15] With

[15]See for this issue on pseudo maximum likelihood estimation in count data models e.g. Gourieroux et. al. (1984a,1984b) or Hausman, Hall, and Griliches (1984).

zeros, zero inflated and hurdle models provide different solutions to the problem of overdispersion due to excess zeros.

Overdispersion, however, might not only be caused by excess zeros but also by heterogeneity. An obvious way to cope with heterogeneity among individuals is to include covariates in the mean function. If the mean still is not perfectly predicted one can additionally include an individual specific random variable to capture unobserved heterogeneity. In the Poisson model e.g. the mean was specified as $E(Y_i) = \mu_i = \exp(X_i\beta)$. Multiplying a gamma distributed density ν with $E(\nu) = 1$ and $Var(\nu) = \alpha$ with the mean, i.e. $E(Y_i) = \exp(X_i\beta)\nu$, and integrating ν out of the Poisson distribution leads to the negative binomial distribution:

(5.11)

$$f_{nb}(Y_i = y_i|\nu, \alpha) = \frac{\Gamma(\alpha^{-1} + y_i)}{\Gamma(\alpha^{-1})\Gamma(y_i + 1)} \left(\frac{\alpha^{-1}}{\alpha^{-1} + \mu_i}\right)^{\alpha^{-1}} \left(\frac{\mu_i}{\mu_i + \alpha^{-1}}\right)^{y_i} \quad \text{with } \alpha > 0$$

and $\Gamma(\cdot)$ denoting the gamma function. The variance is a quadratic function of the mean: $Var(Y_i) = \mu_i + \alpha\mu_i^2$. Hence the distribution in (5.11) is also called NegBin II model.[16] A more general class of models is obtained by assuming that α is no longer constant across all individuals but a function of the mean, i.e. $\alpha_i = \alpha^{-1}\mu_i^{2-p}$. The variance is then given by $Var(Y_i) = \mu_i + \alpha\mu_i^p$. For $p = 1$ one obtains the NegBin I model where the variance is a linear function of the mean.

In order to capture overdispersion due to excess zeros as well as to unobserved heterogeneity and taking the two-step nature of our decision process into account, in the following we will estimate a hurdle model where we assume that both distributions, $f_1(y_i|\theta_1)$ and $f_2(y_i|\theta_2)$, follow a NegBin I distribution. The log-likelihood function for such a model can be obtained by inserting:

$$\begin{aligned} (5.12)\quad f_1(0|\theta_1) &= \left(\frac{\alpha_1^{-1}}{\alpha_1^{-1} + 1}\right)^{\alpha_1^{-1}\mu_{1i}} \\ 1 - f_2(0|\theta_2) &= 1 - \left(\frac{\alpha_2^{-1}}{\alpha_2^{-1} + 1}\right)^{\alpha_2^{-1}\mu_{2i}} \\ f_2(Y_i = y_i|\theta_2) &= \frac{\Gamma(\alpha_2^{-1}\mu_{2i} + y_i)}{\Gamma(\alpha_2^{-1}\mu_{2i})\Gamma(y_i + 1)} \left(\frac{\alpha_2^{-1}\mu_{2i}}{\alpha_2^{-1}\mu_{2i} + \mu_{2i}}\right)^{\alpha_2^{-1}\mu_{2i}} \left(\frac{\mu_{2i}}{\mu_{2i} + \alpha_2^{-1}\mu_{2i}}\right)^{y_i} \end{aligned}$$

into (5.9). Thereby $\mu_{1i} = \exp(X_{1i}\beta_1)$ and $\mu_{2i} = \exp(X_{2i}\beta_2)$ are the mean functions of the two NegBin I processes, respectively. We will compare this model with a

[16]For more details see e.g. Cameron and Trivedi (1998).

Poisson hurdle as well as with the standard representations of the Poisson and the NegBin I model in the following section.[17]

5.5 Estimation Results

In a first step we estimated a standard Poisson model with the following explanatory variables: Proxies for the four Lindbeck/Snower factors including quadratic terms for the share of high skilled and young employees, size of the establishment measured by the number of employees, a dummy variable for East German establishments, industrial sector dummies and measures for competition pressure, business development and export share.[18] Table D.1 in the appendix contains a detailed description of each variable together with some descriptive results. Estimating a standard NegBin I model with the same explanatory variables clearly indicates the presence of overdispersion thus leading to a rejection of the standard Poisson model (see also figure 5.6).

Figure 5.6: Model Specification and Selection Tests

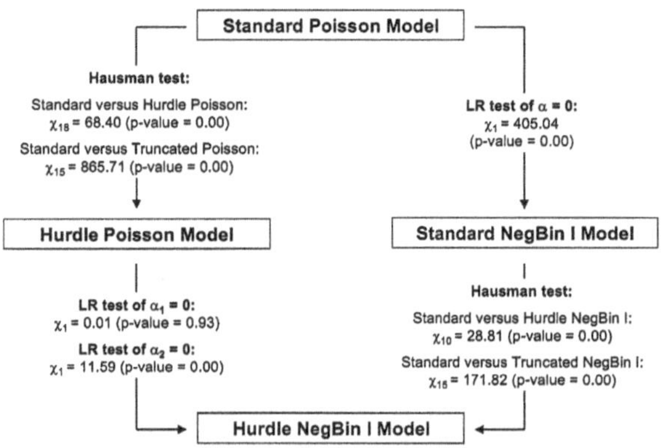

In a second step we estimated a hurdle Poisson and a hurdle NegBin I model. Like already discussed in section two, the first stage decision whether to intro-

[17]Pohlmeier and Ulrich (1995) apply such a hurdle specification to the demand for health care in Germany.

[18]We considered the following industrial sectors: Electricity/Mining, primary industry, investment goods industry, consumer goods industry, construction, commerce, transportation/communications, financing/insurance, services. The likelihood function was programmed in STATA.

duced organizational changes or not is assumed to be determined by the internal Lindbeck/Snower factors and by the external factors competition pressure, business development and export share. The decision about the number of introduced measures on the other hand is assumed to depend only on the Lindbeck/Snower factors. Control variables such as industrial affiliation, size and location of establishments were again included in all models.[19] Conducting Hausman tests reveals that the standard count data models compared to their hurdle counterparts are not appropriate. Comparing finally the hurdle NegBin I with the hurdle Poisson model yields a significant α_2 thus indicating the presence of overdispersion in the truncated part of the distribution. In the following we will therefore only report the estimation results for the hurdle NegBin I and the standard NegBin I model as a benchmark (see table 6.1).

Table 5.1: NegBin I and Hurdle NegBin I Estimation Results

	Standard NegBin I		Hurdle NegBin I		Truncated NegBin I	
Variable	Coefficient	t-Value	Coefficient	t-Value	Coefficient	t-Value
Constant	-0.90	-5.56	–	–	0.22	1.67
$STATE_{98}$	-0.02	-0.89	0.00	0.13	-0.01	-0.61
ICT_{97}	0.52	13.13	0.54	10.83	0.32	7.99
$HIGH_{98}$	1.25	3.37	1.40	3.07	0.46	1.44
$HIGH^2_{98}$	-1.09	-2.14	-1.27	-1.97	-0.61	-1.16
$YOUNG_{98}$	3.32	11.45	3.06	9.02	1.99	5.59
$YOUNG^2_{98}$	-3.31	-12.20	-3.16	-9.85	-1.86	-5.55
$EMPL_{98}$	0.11	5.06	0.23	6.08	0.03	6.06
EAST	-0.02	-0.50	0.11	2.34	-0.17	-5.03
$COMP_{98}$	0.13	4.85	0.14	3.99	–	–
UP_{98}	0.20	5.07	0.22	4.46	–	–
EXP_{97}	0.00	3.88	0.00	2.73	–	–
α	0.46	13.65	0.61	0.01	0.14	3.4
	χ^2-Value	p-Value	χ^2-Value	p-Value	χ^2-Value	p-Value
LR-Industry	73.71	0.00	53.79	0.00	77.95	0.00
LR-Overall	1,010.13	0.00	736.86	0.00	358.65	0.00
Observations	4,090		4,090		5,035	

Several points are worth noting. First of all, likelihood ratio tests on the overall significance reveal a highly significant association of the included covariates with the dependent variable in all estimated models.[20] Comparing the parameters of the

[19]Size of establishment was measured by the number of 1,000 employees.

[20]Note that the intercept term and the dispersion parameter of the hurdle NegBin I model are not

variables included in the hurdle and the truncated model, one can see that there are considerable differences. Location of the establishment e.g. is insignificant in the standard NegBin I model. When differentiating between the qualitative and the quantitative decision, however, it turns out that East German establishments are significantly more likely to introduce organizational changes at all but that they introduce less changes than their West German counterparts.

Regarding the Lindbeck/Snower factors state of technology, in contrast to investment in ICT, turns out to be insignificant in all models, while the share of high skilled employees is significant in the hurdle but less significant in the truncated distribution. Additionally, we found evidence for quadratic effects for the share of high skilled and young employees with peaks at 0.57 (0.55/0.38) for the share of high skilled employees and 0.50 (0.48/0.53) for the share of young employees for the standard (hurdle/truncated) NegBin I model, respectively. An increasing competition pressure, export share or upswing in the expected business development increases the probability to introduce organizational changes.

Since the estimated parameters cannot be interpreted directly, in the following we calculated the relative importance, r_j, of the j-th factor, x_j, given by:

$$r_j = \frac{E(Y|x_j^{max}) - E(Y|x_j^{min})}{\sum_j E(Y|x_j^{max}) - E(Y|x_j^{min})}. \tag{5.13}$$

Thereby x_j^{max} (x_j^{min}) denotes the value of x_j for which the dependent variable takes on its maximum (minimum) value. We estimated the relative importance of the Lindbeck/Snower factors for an establishment located in West Germany, operating in the investment goods industry and given that all other covariates take on their mean values. Table 5.2 contains the results as well as the total variation in the dependent variable due to a simultaneous variation of the Lindbeck/Snower factors. Standard errors were bootstrapped with 1,000 replications.[21]

Versatile capital measured by the state of technology has the least importance and turns out to be insignificant in all estimated models. The remaining factors play different roles in the qualitative decision to introduce organizational changes and the quantitative decision about the number of introduced measures. In the former, worker tastes measured by the share of younger employees and investment

separately identified. See e.g. Pohlmeier and Ulrich (1995) for a formal proof.

[21]The results are quite robust for different industry sectors and location. Table D.2 in the appendix contains the relative importance of the Lindbeck/Snower factors for a representative West German establishment in the service sector as well as for East German establishments in the investment goods industry and service sector. Again all remaining variables are assumed to take on their mean values.

Table 5.2: Relative Importance of the Lindbeck/Snower Factors

Factor	Standard NegBin I Coefficient	Standard NegBin I t-Value	Hurdle NegBin I Coefficient	Hurdle NegBin I t-Value	Truncated NegBin I Coefficient	Truncated NegBin I t-Value
Versatile capital	0.08	1.52	0.07	1.43	0.19	1.49
Investment ICT	0.34	5.94	0.30	5.84	0.28	3.49
Human capital	0.21	1.89	0.16	1.46	0.28	1.80
Worker tastes	0.37	6.21	0.47	6.40	0.25	2.82
Total ΔY	2.69	5.94	0.64	6.09	1.72	2.67

in ICT play the dominant role, while in the latter case investment in ICT, the share of high skilled employees and the share of younger employees are equally important. Estimating a standard count data model will thus mask complex structures which can only be disclosed by a hurdle formulation.

Varying the Lindbeck/Snower factors has a numerical considerable and significant impact on the dependent variable: The probability to introduce organizational changes increases by about 0.64 and given a positive first stage decision, the average number of organizational changes by 2.67. An interesting question concerns the joint contribution and relative importance of the Lindbeck/Snower factors as compared to other control variables, namely industrial affiliation, location and size of the establishment. To this aim we estimated two separate models and conducted a variety of tests to discriminate between them. In the first model, M_{LS}, only the Lindbeck/Snower factors were included while the second, M_{rest}, contains only the remaining covariates.

One way to compare two non-nested models consists in computing familiar model selection criteria like Akaike's (AIC) or Schwarz's Bayesian information criteria (SBC).[22] Additionally, we also conducted the likelihood ratio test for non-nested models proposed by Vuong (1989) and a nonparametric counterpart described in Clarke (2001,2003a,2003b).

The basic idea of the Vuong test is that the average log-likelihoods of the two models M_{LS} and M_{rest} should be equal if both groups of covariates have the same relative importance regarding the dependent variable. If however the Lindbeck/Snower factors are more (less) important than the remaining control variables the difference between the two log-likelihoods will be significantly greater

[22] See Akaike (1973) and Schwarz (1978).

(smaller) than zero. Vuong (1989) derived a suitable test statistic and showed that it follows a standard normal distribution.

One drawback of the Vuong approach is that this test is only asymptotically valid. An alternative is a nonparametric paired sign test applied to the differences in the individual log-likelihoods from M_{LS} and M_{rest}, respectively. The test determines whether the median log-likelihood ratio is statistically different from zero or not. If model M_{LS} (M_{rest}) is closer to the real *dgp* than more (less) than half of the individual log-likelihood ratios should be greater than zero, indicated by $f_{LS} > f_{rest}$ ($f_{LS} < f_{rest}$).

Table 5.3: Joint Contribution of the Lindbeck/Snower Factors

	Standard NegBin I		Hurdle NegBin I		Truncated NegBin I	
Model	M_{LS}	M_{rest}	M_{LS}	M_{rest}	M_{LS}	M_{rest}
AIC	15,605	12,434	6,146	5,169	9,199	6,956
SBC	15,644	12,516	6,185	5,258	9,238	7,038
$\ln L$	-7,797	-6,204	-3,067	-2,571	-4,593	-3,465
	LR-Value	*p*-Value	*LR*-Value	*p*-Value	*LR*-Value	*p*-Value
Vuong-test	-101.74	0.00	-35.03	0.00	-267.61	0.00
Corrected	-73.25	0.00	-2.38	0.01	-239.12	0.00
	$f_{LS} > f_{rest}$	$f_{LS} < f_{rest}$	$f_{LS} > f_{rest}$	$f_{LS} < f_{rest}$	$f_{LS} > f_{rest}$	$f_{LS} < f_{rest}$
Clarke-test	0.00	1.00	0.00	1.00	1.00	0.00
Corrected	0.00	1.00	0.00	1.00	0.00	1.00

Table 5.3 contains the appropriate figures. All information criteria point to the view that the remaining covariates included in the regression have a larger contribution regarding the dependent variable than the Lindbeck/Snower factors since both information criteria are smaller for M_{rest}. The Vuong test reveals that these differences are also statistically significant. Correcting the Vuong test for the appropriate degrees of freedom confirms this impression.

A crosscheck using the nonparametric sign test yields the same conclusion. The only exception is found for the truncated NegBin I model. However, correcting the Clarke test for the number of included parameters reverses the result. It thus seems that albeit individually and jointly significant, compared to other control variables the Lindbeck/Snower factors play only a minor role in explaining organizational changes.

5.6 Summary and Conclusions

The aim of this chapter was to undertake an empirical analysis of the recent shifts within establishments from a 'Tayloristic' to a 'Holistic' organization of work. In their seminal work Lindbeck and Snower (2000) identified four major forces driving these changes: Investment in more versatile equipment and information and communication technology, advances in human capital and changing worker preferences in favor of more varied jobs.

Using an unique linked employer-employee dataset and hurdle count data models we were able to show that these forces have a complex impact on organizational changes. Regarding the decision to introduce organizational changes worker preferences and investment in ICT play the dominant role while for the quantitative decision on how many measures to introduce all factors except for versatile equipment are equally important. Compared with other control variables like industrial affiliation, region and size, however, it turns out that the Lindbeck/Snower factors play only a minor role in explaining organizational changes.

Organizational changes have a number of considerable economic impacts. Several studies indicate that they increase the productivity and performance of establishments. In the previous chapter we have shown that they additionally exhibit an innovation enhancing effect. However, organizational changes have also serious implications for the labor market. The results of the following chapters and existing empirical studies indicate that organizational changes increase the wages and demand for high skilled employees and are therefore regarded as another 'typical suspect' besides international trade and technological change in explaining wage inequality and rising unemployment especially among low skilled employees.[23] It is therefore imperative to learn more about the process and factors driving organizational changes. This chapter was intended as a first step to this aim.

[23]See especially Snower (1999).

Part III

Consequences of the Innovation Process

Chapter 6

Compensation or Displacement: Skill Biased Technological and Organizational Change?

6.1 Introduction

The first part of this thesis was concerned with an analysis of the determinants of the innovation propensity of German establishments. Once triggered, however, it is also important to take care about the consequences of technological change. In this context the last 30 years have brought growing evidence for an increasing labor demand for high skilled and a deterioration of the labor market position of less skilled employees.

As already discussed in the first chapter of this thesis, in countries with a more flexible wage structure, like the United States and the United Kingdom, this increasing demand for high skilled employees manifests itself in a widening of the distribution of wages and an increasing income inequality.[1] On the other hand, countries with more rigid wages, like most of the European countries, have to face increasing unemployment rates for unskilled workers. The European unemployment problem is thus seen to be caused by the same factors that are responsible for the increasing inequality of income in the U.S. and U.K.[2]

Economists offer three possible explanations for the increasing demand for high skilled employees.[3] One consists in an increasing international trade, often paraphrased with the notion of globalization. According to this view, inter-industry

[1]See e.g. Snower (1999).

[2]Or using the words of Krugman (1994): 'The European unemployment problem and the US inequality problem are two sides of the same coin.'

[3]See also chapter one of this thesis.

trade between the capital and skill abundant 'North' and the labor abundant 'South' will decrease the labor demand for low skilled employees in the northern countries.[4]

Another widespread explanation for an increasing labor demand for high skilled employees is a skill biased technological change, i.e. a technological change which favors high skilled employees. According to this view, technological progress that manifests itself in increasing investments in information technology or in new and more complex product and process innovations requires a higher qualified staff.[5] There is, however, a further explanation that has received less attention until now. Establishments recently introduced numerous organizational changes away from the traditional 'Taylorist' organization characterized by a centralization of authority. Examples include flattening of hierarchical levels or the introduction of more flexible and responsible workplaces. These organizational changes can be expected to require a higher level of human capital and therefore might be skill biased, too.[6]

The purpose of this chapter is to analyze the interdependencies between labor demand for high and low skilled employees, innovation activities and organizational changes. To meet this aim we will use the 1998 wave of the IAB Establishment Panel, a large panel dataset collected by the German Federal Employment Office ('Bundesagentur für Arbeit') which focuses on the demand side situation of the labor market.

The 1998 wave allows us to distinguish between different qualification groups: Blue and white collar employees for simple and qualified jobs. Additionally it contains information about radical and incremental product innovations and organizational changes introduced by the establishments like the ones mentioned above.

Since some of the considered variables are qualitative and presumably jointly dependent, we have to utilize a mixed simultaneous equation model. Thereby two different approaches can be distinguished: Type I models which are defined in terms of the latent variables and type II models with endogenous dummy variables appearing as additional explanatory variables. The structural form parameters of

---

[4]See Berman and Griliches (1994) and Berman, Bound, and Machin (1998), who have conducted shift share-analysis for the U.S. and OECD economies, and Falk and Seim (2001), who have done the same for the German service sector.

[5]See Bartel and Lichtenberg (1987), Krueger (1993), Autor, Katz, and Krueger (1998).

[6]See Crepon, Duguet, and Mairesse (1998), Bresnahan, Brynjolfsson, and Hitt (2002) or Caroli and Reenen (2001).

type II models have to fulfill additional restrictions to ensure the coherency of this model. We will apply both models and pay special attention to the consequences arising from the parameter restrictions for type II models.

The subsequent considerations proceed as follows: In the second section we will present some theoretical considerations about the determinants of labor demand, innovation activities and organizational changes. Section three of this chapter deals with the econometric specification and estimation of mixed simultaneous equation models. Section four introduces briefly the dataset used and presents first descriptive evidence while section five presents the estimation results. Finally, a summary and some concluding remarks are given in section six.

6.2 Theoretical Considerations

The aim of this section is to give a brief overview of the theoretical expected determinants of labor demand, innovation decisions and organizational changes. Having identified potential explanatory variables and functional forms, we will consider the appropriate econometric modelling in the next section.

Labor Demand

We start with deriving factor demand equations by assuming the following general cost function:[7]

$$C = f(w_{high}, w_{low}, w_{app}, K, Y) \tag{6.1}$$

where

$$\begin{aligned} C &= \text{Total cost} \\ w_{high} &= \text{Wage rate for high skilled employees} \\ w_{low} &= \text{Wage rate for low skilled employees} \\ w_{app} &= \text{Wage rate for apprentices} \\ K &= \text{Capital stock} \\ Y &= \text{Output.} \end{aligned}$$

We assume that there are three variable input factors, namely employees with high and low education, respectively, L_{high} and L_{low}, apprentices, L_{app}, and one

[7]See Bartel and Lichtenberg (1987).

quasi fixed input factor, which is capital stock in 1998, K. The group of high skilled employees consists of skilled workers and white collar workers for qualified tasks, whereas low skilled employees include un- and semiskilled blue collar and white collar workers as well as officials for simple tasks.

If we define Z as a column vector consisting of K and Y, (6.1) can also be written more conveniently as:

$$C = f(w_{high}, w_{low}, w_{app}, Z). \tag{6.2}$$

There are several cost functions which can be used to approximate (6.1).[8] The most flexible is the translog cost function leading to the following specification:

$$\begin{aligned} \ln C &= \ln \alpha_0 + \sum_{i \in I} \alpha_i \ln w_i + \frac{1}{2} \sum_{i \in I} \sum_{j \in I} \alpha_{ij} \ln w_i \ln w_j \\ &+ \sum_{i=1}^{2} \beta_i \ln Z_i + \frac{1}{2} \sum_{i=1}^{2} \sum_{j=1}^{2} \beta_{ij} \ln Z_i \ln Z_j \\ &+ \frac{1}{2} \sum_{i \in I} \sum_{j=1}^{2} \gamma_{ij} \ln w_i \ln Z_j \end{aligned} \tag{6.3}$$

with $I = \{high, low, app\}$, $\alpha_{ij} = \alpha_{ji}$, $\beta_{ij} = \beta_{ji}$ and $\gamma_{ij} = \gamma_{ji}$. For (6.4) to be homogenous of degree 1 in prices, the following restrictions have to be imposed:

$$\sum_i \alpha_i = 1 \text{ and } \sum_i \alpha_{ij} = \sum_j \alpha_{ji} = \sum_i \gamma_{ij} = 0. \tag{6.4}$$

Differentiating (6.4) with respect to the factor prices and applying Shephard´s lemma:

$$\frac{\partial \ln C}{\partial \ln w_i} = \frac{w_i L_i}{C} = S_i \quad \text{for } i = high, low \text{ and } app$$

yields the following cost share equations:

$$S_i = \alpha_i + \frac{1}{2} \sum_{j \in I} \alpha_{ij} \ln w_j + \frac{1}{2} \sum_{j=1}^{2} \gamma_{ij} \ln Z_j \tag{6.5}$$

for $i = high, low$ and app. Since the three share equations sum up to one, we will skip the share equation for apprentices in the following. Imposing the restrictions

[8] See e.g. Hamermesh (1986, 1993).

in (6.4) we arrive at the following factor demand equations:

$$(6.6) \qquad S_{high} = \alpha_{high} + \sum_{i \in \{high,low\}} \alpha_{high,i} \ln \frac{w_{high}}{w_{app}} + \gamma_{high,i} \ln \frac{K}{Y}$$

$$(6.7) \qquad S_{low} = \alpha_{low} + \sum_{i \in \{high,low\}} \alpha_{low,i} \ln \frac{w_{low}}{w_{app}} + \gamma_{low,i} \ln \frac{K}{Y}.$$

Since the IAB Establishment Panel does not contain any information about wages for different skill groups, we will approximate the cost share for the i-th skill group with their employment share.[9] Furthermore assuming that the relative wages $\frac{w_i}{w_{app}}$ will not vary across industries, we can capture this effect using industry dummies.[10] The physical capital stock will be approximated by the investments of the last year.

These theoretically founded employment share equations will be used further on for the empirical implementation. Additionally to the industry dummies we will also include other control variables that might affect the labor demand. Possible candidates are the business development of the past year, investments in communication and information technology, competition pressure, state of technology and regional dummies. Since our main interest lies in the employment effects of product innovations and organizational changes, corresponding dummy variables will also be taken into account. Table E.1 in the appendix contains a description of the variables used in the analysis.

Since we have not included factor prices in the equations, no symmetry or other restrictions have to be considered and hence the augmented versions of (6.6) and (6.7) could be estimated separately using OLS. Table E.4 and E.5 in the appendix contain the so obtained estimation results for high and low skilled employees. We will discuss these in section five.

One of the main problems with this proceeding, however, is the potential endogeneity of the innovation and organizational change dummies in the labor demand equations. Introduced product innovations and organizational changes clearly depend on a number of decisions made by the establishments and therefore should be modelled in a simultaneous equation framework. In the next two sections we will discuss possible determinants of innovation decisions and organizational changes

[9] See e.g. Berman and Griliches (1994) or Falk and Seim (2001) who proceed in the same manner. In the next chapter we will relax this assumption using a linked employer employee dataset.

[10] We considered 16 different industry sectors in our estimation. See the appendix for details. Excluding the aggriculture/forestry, government/administration and the non profit organization sector does not alter the estimation results.

in greater detail before turning to the problem of simultaneity. We will then compare the estimation results obtained by applying OLS to (6.6) and (6.7) with more appropriate methods which take a possible simultaneous bias into account.

Innovation Decision

Technological change can be roughly divided into process and product innovations (see chapter one of this thesis and also figure 6.1). The first refers to any changes that enable establishments to produce a given output at lower costs, while the latter means the introduction of new products and services. Process innovations can be further distinguished as to whether they come along with investments in new capital goods (embodied process innovation), e.g. investments in information and communication technology, or whether they reflects an increase in the production efficiency (disembodied process innovation). Product innovation can be categorized into improvement of existing products and the introduction of products that are new for the establishment (incremental product innovation) or new for the market (radical product innovation).

Figure 6.1: Classification of Innovations

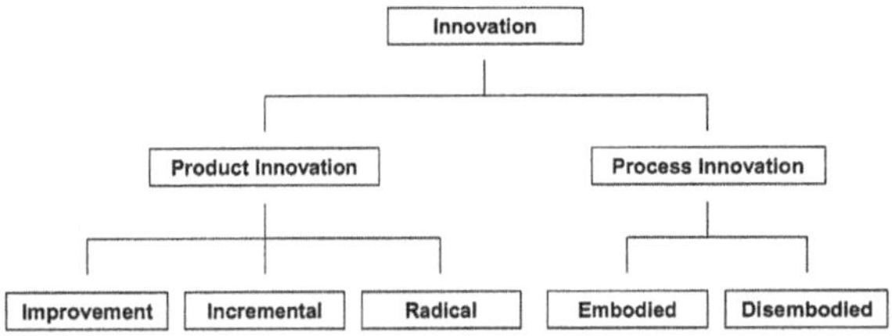

It is widely believed that product innovations, in opposition to process innovations, have a positive employment effect. Especially for radical product innovations, i.e. products which are not close substitutes to existing ones, Katsoulacos (1986) has shown that the positive welfare effect which stems from the increase of the workers' utility from consumption leads to an increase in the total employment. The theoretical literature is, however, relatively scarce with regard to the employment effects for different qualification groups of employees. This theoretical ambiguity makes an empirical assessment necessary.

Our dataset allows us to distinguish between these various forms of product innovations introduced by the establishments during the years 1996 and 1997:

Improvement, incremental and radical. Since there are high positive correlations between them (see the next section for some descriptive results), we will combine them to one single dummy variable, $INNO$, which equals one if at least one of the innovation forms is realized and zero otherwise.

Embodied process innovations are in part captured by the total net investments, especially investments in information and communication technology, ICT. Our dataset, however, allows us a much broader view of process innovation. While the use of computers has been in the center of the debate about the employment consequences of technological change, it has been recognized recently that exploiting full advantage from computerization requires a redesign of the organization of workplaces, too. Technological change and changes in the organizational structure of workplaces are closely related.[11] We will take this into account by explicitly considering organizational changes as an integral part of disembodied process innovations.

With regard to the product innovation decision, we assume a latent threshold model:

$$INNO = \begin{cases} 1 & \text{if } INNO^* \geq 0 \\ 0 & \text{otherwise} \end{cases} \tag{6.8}$$

where $INNO^*$ can be interpreted as the propensity of an establishment to innovate, depending linearly on a set of explanatory variables, i.e. $INNO^* = \beta'_{inno}x_{inno} + u_{inno}$. For the error term we assume $u_{inno} \sim IIND(0, \sigma_{inno})$. In the following we will discuss some potential determinants of the innovation decision. Thereby a distinction in firm specific and market specific characteristics proves to be useful which was already introduced in chapter one of this thesis.

Firm specific factors include factors like the qualification structure of employees. The human factor is without doubt one of the major factors for the innovation propensity of establishments. We will make use of the following indicator to measure the human capital of an establishment: Share of high qualified white and blue collar employees in 1996. Additionally, we will take into account whether establishments possess departments for both research and development (R&D) and marketing research, respectively. One might also think that certain organizational environments like e.g. flatter and flexible hierarchies, are more beneficial for innovation activities than others. We will test this hypothesis by including an

[11] See e.g. Bell (1996) and Bresnahan, Brynjolfsson, and Hitt (2002).

appropriate dummy variable in the innovation equation. Another possible determinant which will be taken into account is the effect of profit participation plans for employees.

Establishments that do not execute R&D themselves can capture positive spillover effects by cooperating with other institutions, e.g. customers, suppliers, competitors or universities. We will take this possibility into account by including the dummy variable $COOP$ which equals one, if an establishment is engaged in a R&D cooperation with one of these institutions. Additionally, we will consider the total investments in 1997 as a proxy for the capital stock, investments in information and communication technology in 1995 and the state of the technology in 1996 as other possible factors for the innovation activities.

Possible push factors coming from the market side will be captured by the business development for the year 1997 and the competition pressure perceived by the establishments in 1998. Unfortunately, our dataset does not contain any other information about the market structure, e.g. market power and diversification of single establishments or industrial concentration.

Size of the establishment, which is traditionally considered as another important determinant of innovation activities, will be taken into account using dummy variables for different size classes. In addition industry and regional dummies as well as the legal form of the establishment will also be included. Table E.2 in the appendix contains a detailed description of the variables used in the analysis.

Again one obvious idea would be to estimate equation (6.8) directly using standard maximum likelihood methods. The major drawback of this proceeding, however, consists in a potential endogeneity bias regarding the organizational change dummy. Table E.6 in the appendix contains the standard ML-estimation results in the column labelled ML. In section five these results will be contrasted to those obtained by using more appropriate econometric models which take the potential endogeneity of the dummy regressors into account.

Organizational Changes

It is nowadays generally accepted that establishments with a higher degree of functional flexibility are in a better position to react and adapt promptly to changing economic conditions and new opportunities. This flexibility is especially necessary in an economic environment with increasing competition and pressure to adjust to changing circumstances.[12] In chapter five we have undertaken an econometric analysis to reveal the forces driving this process which can be characterized by a trend away from a centralized, so called 'Taylorist' organization. Instead, more and more establishments introduced decentralized and flexible workplaces with more responsibilities for the individual worker and fewer hierarchical levels.

Clearly there are complementarities between the internal organization of an establishment and the above mentioned embodied process innovations, like e.g. investments in ICT.[13] Better communication possibilities associated with ICT also require appropriate organizational changes to fully exploit the advantages of computerization. In this sense technological change and changes in the organizational structure of workplaces are closely related. Similarly, product innovations might have an impact on introduced organizational changes, too.

In the following we will consider whether establishments have undertaken one of the following organizational changes during the years 1996 and 1997 which have been also analyzed in chapter five:

- Reorganization of departments and functional areas,
- Decentralization of responsibilities and decision-authority,
- Introduction of group-workplaces with own responsibilities or
- Introduction of units with own costing and result calculation.

Since there are high positive correlations between these changes (see the next section for some descriptive results), we will combine them to one single dummy variable, $ORGA$, which equals one if at least one of the organizational changes is realized and zero otherwise.

The decision about the introduction of such organizational changes depends on a number of factors. One obvious one is the qualification structure of the employees. More qualified employees are more able to analyze the growing information flood caused by flatter hierarchies and to exploit the scope offered by workplaces with more responsibilities. Thus we expect establishments with a higher qualified staff to have a higher propensity to introduce organizational changes.

Another main factor that might have a positive impact on the introduction of organizational changes is a growing competition pressure since it increases the

[12]Lundvall and Kristensen (1997).

[13]See Bresnahan, Brynjolfsson, and Hitt (2002).

gains of flexibility and decentralization. We will capture this effect using assessments of the establishments regarding the competition pressure as well as the business development.

We have already mentioned the expected complementarities between embodied process innovations as measured by investments in ICT and organizational changes. Additional explanatory variables, which will be taken into account, therefore consist of investments, especially in ICT. To consider possible effects of product innovations on the propensity to introduce organizational changes, we will also include the dummy variable $INNO$.

To capture other firm characteristics, industrial and regional as well as size dummies will also be included. Machin and Wadhwani (1991) pointed out that workers councils might have an impact on the introduction of organizational changes. We will therefore include an appropriate dummy variable as well as variables indicating the legal form of the establishment. Table E.3 in the appendix contains a description of the variables used for estimating the following organizational change equation:

$$ORGA = \begin{cases} 1 & \text{if } ORGA^* = \beta'_{orga} x_{orga} + u_{orga} \geq 0 \\ 0 & \text{otherwise} \end{cases} \tag{6.9}$$

with $u_{orga} \sim IIND(0, \sigma_{orga})$.

Table E.7 in the appendix contains the estimation results obtained by employing standard ML-estimation methods in the column labelled ML.[14] In the following we will examine if these results deviate from the estimation results obtained from a mixed simultaneous equation model.

6.3 Econometric Modelling

Collecting the equations for the share of high and low skilled employees, innovation decision and introduced organizational changes, yields the following simultaneous equation model:

$$S_{high} = \gamma_{13} INNO + \gamma_{14} ORGA + \beta'_{high} x_{high} + u_{high} \tag{6.10}$$
$$S_{low} = \gamma_{23} INNO + \gamma_{24} ORGA + \beta'_{low} x_{low} + u_{low} \tag{6.11}$$
$$INNO^* = \gamma_{34} ORGA + \beta'_{inno} x_{inno} + u_{inno} \tag{6.12}$$
$$ORGA^* = \gamma_{43} INNO + \beta'_{orga} x_{orga} + u_{orga} \tag{6.13}$$

[14]Section five will go further into detail concerning the interpretation of the estimation results.

where the observable dummy variables are linked with their latent counterpart via the following threshold models:

$$(6.14)\quad INNO = \begin{cases} 1 & \text{if } INNO^* \geq 0 \\ 0 & \text{otherwise} \end{cases} \quad \text{and} \quad ORGA = \begin{cases} 1 & \text{if } ORGA^* \geq 0 \\ 0 & \text{otherwise.} \end{cases}$$

Since some of the endogenous explanatory variables are qualitative, models like the one in (6.10) - (6.14) are called mixed simultaneous equation models.

Classical simultaneous equation models in which all encountered variables are continuous are well known to econometricians and have been successfully applied to a variety of problems, mainly macroeconomic ones involving longitudinal data sets.

However, the rapid growth and availability of microeconomic data sets have recently also raised the interest in discrete data and appropriate econometric models since individual decisions and choices reflected in such data sets are often not made over a continuum but over a limited number of possibilities and alternatives.

Another already mentioned problem besides the qualitative nature of some variables is due to the fact that some explanatory variables are the outcome of other decisions by the individuals and therefore have to be modelled simultaneously.

Mixed simultaneous econometric models (MSEM) can be used when working with data sets involving both quantitative and qualitative jointly dependent variables. The field of applications is broad ranging from consumer analysis, labor economics to questions coming from financial analysis. Since MSEM might be computational demanding and feature additional identification, consistency and coherency problems, the applications are rather scarce.[15]

By modelling a simultaneous model with endogenous dummy, ordered and/or truncated variables, two different approaches can be distinguished:[16]

[15]Some examples are Lee (1978) and Schmidt (1978): Analysis of the impact of labor unions on wage rates of workers, Nelson and Olson (1978): Analyzing the determinants of decisions about post secondary vocational school training, Waldman (1981): Analyzing the interrelation between work, leisure and schooling time decision, Ransom (1987): Analyzing family labor supply and consumption decision, Pohlmeier (1989): Analysis of market structure and innovation activities, Browning (1992): Children and household economic behavior, Blundell and Smith (1993): Joint decision model for married women´s hours of work and household income and Li (1998): Joint determination of a financially distressed high yield company´s length of time in default and its likelihood of seeking bankruptcy protection.

[16]See e.g. Blundell and Smith(1993,1994) or Wilde (1999). Besides type I and type II models so called log-linear models have been proposed in the literature. Since these models, however, can not be

- Type I Models: Latent variable formulation or

- Type II Models: Dummy variable formulation.

Type I models are basically simultaneous equation models defined in terms of the underlying latent variables. Let y_i^* be a $(G \times 1)$ vector which contains the latent variables for the i-th individual, x_i a $(K \times 1)$ vector with explanatory variables, u_i a $(G \times 1)$ vector with stochastic disturbances, Γ a $(G \times G)$ and B a $(G \times K)$ matrix containing the structural form parameters.

A mixed simultaneous type I model can then be written as:

$$\Gamma^* y_i^* + B x_i = u_i. \tag{6.15}$$

Since (6.15) corresponds to a standard simultaneous equation model defined in terms of y_i^*, an explicit and unique reduced form for (6.15) exists which can be used to estimate the reduced form parameters. The only additional problem which arises is due to the fact that not all of the endogenous variables are directly observable thus making some additional identification conditions necessary.

Whereas type I models are defined in terms of the latent variables, mixed simultaneous type II models might also contain the observable dependent variables as regressors. Using the notation above, type II models can be written as:

$$\Gamma^* y_i^* + \Gamma y_i + B x_i = u_i. \tag{6.16}$$

Since in (6.16) not only y_i^* but also y_i appears, no explicit reduced form for this model exists. Additional to the identification criterion, the structural form parameters in (6.16) have to fulfill the so called coherency condition which ensures that for every admissible value of x_i, u_i, B, Γ^*, and Γ a unique value of y_i is associated.

This coherency condition must be distinguished from the identification condition.[17] The former ensures the existence and uniqueness of the reduced form whereas the latter ensures the uniqueness of the parameters of the structural form. Identification therefore presupposes the coherency of a model.

---

interpreted in a structural manner, we will not consider them here. For a detailed discussion see e.g. Maddala (1983).

[17]See e.g. Gourieroux, Laffont and Monfort (1980)

At this point it should be stated that type I and type II models are nonnested alternatives. This follows from the fact that simply setting Γ equal to zero in (6.16) does not yield a type I model since type II models require additional parameter restrictions due to the coherency problem.[18]

The decision when to apply which model should be guided by economic considerations. In type I models, individual behavior is completely described by the latent variables and the censoring process simply acts as a constraint on the information available to the econometrician whereas for type II models the observability rule also constraints the agent´s set of choices.

Bresnahan and Smith (1993) consider a model describing household labor supply and consumption behavior. If the hours of work are only observable in grouped form, e.g. full and part time, although the individual has the choice over a scope of hours of work, the appropriate model would be a type I model. In this case consumption would be a function of the underlying latent hours of work and not of the observable hours of work.

If on the other hand individuals only have the choice between full and part time work, not the latent but the observable dummy variable would have an impact on the consumption decision of households and thus a type II model would be appropriate. Type II models are equivalent to the so called hybrid model with structural shift proposed by Heckman (1978). In this model both the latent and the observable endogenous variable appear as regressors in one equation. Another special case of type II models are switching regression models described in Gourieroux, Laffont and Monfort (1980).

Since in our empirical case of the employment effects of innovations and organizational changes there are no clear theoretical suggestions whether only the actual observable decisions have an impact on other variables, as in the above formulated model (6.10) - (6.14), or also the latent propensity to innovate or introduce organizational changes, in the following we will estimate both models and test which one is appropriate. In the next two subsections we will take a closer look at both of them.

---

[18]See Blundell and Smith (1993).

6.3.1 Mixed Simultaneous Equation Models of Type I

Identification

Let us write down the structural form of the type I mixed simultaneous equation model using matrix notation:

$$\Gamma^* y_i^* + Bx_i = u_i \tag{6.17}$$

where

$$\Gamma^* = \begin{pmatrix} 1 & 0 & -\gamma_{13} & -\gamma_{14} \\ 0 & 1 & -\gamma_{23} & -\gamma_{24} \\ 0 & 0 & 1 & -\gamma_{34} \\ 0 & 0 & -\gamma_{43} & 1 \end{pmatrix} \quad y_i^* = \begin{pmatrix} S_{high} \\ S_{low} \\ INNO^* \\ ORGA^* \end{pmatrix}_i \quad B = \begin{pmatrix} -\beta_{high} \\ -\beta_{low} \\ -\beta_{inno} \\ -\beta_{orga} \end{pmatrix} \quad u_i = \begin{pmatrix} u_{high} \\ u_{low} \\ u_{inno} \\ u_{orga} \end{pmatrix}_i,$$

x_i is a $(K \times 1)$ vector containing all exogenous variables and the β's are $(1 \times K)$ vectors containing the structural form parameters. For the vector u_i it is assumed that $E(u_i) = 0, \forall i = 1, \ldots, N$ and that there are no correlations between different individuals, i.e. $E(u_i u_j) = 0, \forall i \neq j$. The variance covariance matrix of the structural form residual for the i-th individual is given by $E(u_i u_i') = \Sigma$.

The dependency structure between the jointly endogenous variables can be summarized as in figure 6.2. We assume that the propensities to innovate and to introduce organizational changes during the years 1996 and 1997 might have an impact on the labor demand for high and low skilled employees in 1998, whereas the innovation decision and the decision to introduce organizational changes are not affected by the actual employment situation but the situation prevailing in 1996. On the other hand, we allow simultaneous relations between the propensity to innovate and the propensity to introduce organizational changes.

Figure 6.2: Type I Dependency Structure and Econometric Models

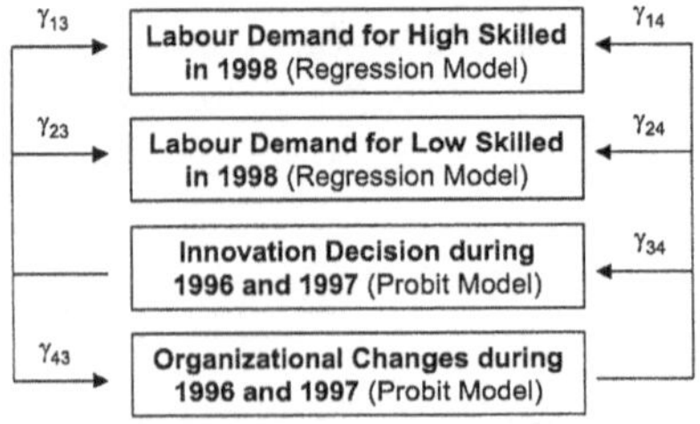

The model in (6.17) represents a classical simultaneous equation model defined in terms of the latent variables. Therefore the usual identification problems arise which are solved by excluding from every equation at least as much exogenous variables as the number of included endogenous variables, i.e. imposing a number of zero restrictions on the $(1 \times K)$ vectors β_{high}, β_{low}, β_{inno} and β_{orga}.

However, identification problems arise not only due to the fact that the number of structural form parameters exceeds the reduced form parameters. Since some of the latent variables are not observable, additional identification problems arise even for the reduced form parameters of the two Probit equations. To demonstrate this let us first of all derive the reduced form of (6.17) which is given by:

$$y_i^* = \Pi x_i + v_i \tag{6.18}$$

with

$$\Pi \equiv -\Gamma^{*-1}B, \quad v_i \equiv \Gamma^{*-1}u_i \quad \text{and} \quad E(v_i v_i') = \Omega. \tag{6.19}$$

The parameters of the structural form are hence linked with the parameters of the reduced form via the following equation:

$$\Gamma^* \Pi + B = 0. \tag{6.20}$$

Assuming that the usual identification criteria are met, the reduced form parameters of the first two regression equations are exactly identified whereas the reduced form parameters of the Probit equations are only estimable up to a scaling parameter $1/\lambda_g$ with λ_g being the variance of the g-th reduced form Probit equation. The same holds true for the covariance of the reduced form residuals between two equations involving at least one Probit equation which is also identified only up to a scaling parameter $1/\lambda_g$.

Due to the fact that not all variables are directly observable, the structural form parameters are also only identified up to a scaling parameter. In order to identify them, in the following we will restrict the variances of the Probit equations to one, i.e. set $\lambda_g = 1$ for the g-th reduced form Probit equation.

Estimation of Type I Models

We now turn to the question of how to estimate the parameters of a MSEM of type I. Two approaches can be distinguished: System estimation and single equation estimation procedures.

System estimation procedures are full information procedures using all information contained in the different equations. They are computationally more burdensome since they require the evaluation of a n-fold integral in the case of a MSEM with n Probit equations.

This problem can be avoided by the use of single equation estimation procedures which are in fact two stage estimators. In a first step the parameters of the reduced form are estimated by means of standard methods. The so obtained parameters are then used in a second step to estimate the parameters of the structural form.

There are several procedures which were proposed in this context.[19] We will apply the two most common ones, namely the two stage estimators proposed by Amemiya (1974, 1979) and Nelson and Olson (1978).

The Amemiya Estimation Procedure Amemiya (1978, 1979) suggests after estimating the reduced form parameters in Π by standard methods, to use equation (6.20) to estimate the structural form parameters. With Γ_g and B_g as the g-th row of Γ and B respectively, for the g-th equation we have:

$$\Gamma_g \Pi = B_g. \tag{6.21}$$

We further define:

γ_g $(1 \times r)$ row vector of Γ_g which contains the parameters of the r endogenous variables included in the g-th equation,

β_g $(1 \times s)$ row vector of B_g which contains the parameters of the s exogenous variables included in the g-th structural form equation,

π_g $(1 \times K)$ row vector which contains the reduced form parameters of the K exogenous variables in the g-th reduced form equation,

$\underline{\Pi}_g$ submatrix of Π containing the parameters to be multiplied with γ_g and

J_g selection matrix such that $J_g \beta_g' = B_g'$.

After transposing, equation (6.21) can be written as:

$$\pi_g' = \underline{\Pi}_g' \gamma_g' + J_g \beta_g'. \tag{6.22}$$

[19] See Blundell and Smith (1993, 1994) for an overview and Lee (1981) for a discussion of the asymptotic properties of different estimation methods.

The structural form parameters of the g-th equation can then be estimated by substituting the π-parameters in (6.22) by their first step estimates $\hat{\pi}$ and computing the OLS or GLS estimators for γ_g' and β_g'.

The Nelson-Olson Estimation Procedure Nelson and Olson (1978) suggest to estimate in a first step the reduced form parameters by standard methods. These parameters can be used to get predictions for the latent variables according to:

$$\hat{y}_i^* = \hat{\Pi} x_i. \tag{6.23}$$

In a second step the predicted values for y_i^* are used to obtain the parameters of the structural form by inserting the $\hat{y}_i^*$ into the structural form equations and applying standard estimation methods to obtain estimates for Γ and B.

6.3.2 Mixed Simultaneous Equation Models of Type II

Coherency Conditions for the Structural and Semireduced Form

For the following considerations we will use the notation introduced in section 6.3.1 A mixed simultaneous type II model for the i-th individual can then be written as:

$$\Gamma^* y_i^* + \Gamma y_i + B x_i = u_i. \tag{6.24}$$

In contrast to type I models we permit that not only the latent variables contained in y_i^* are considered in the structural model but also some of their observable counterparts in y_i. One consequence of this inclusion is the nonexistence of an explicit reduced form. By premultiplying (6.24) with the inverse of Γ^*, however, a semireduced form can be derived:

$$y_i^* = \widetilde{\Gamma} y_i + \widetilde{B} x_i + \widetilde{u}_i \tag{6.25}$$

with $\widetilde{\Gamma} = -{\Gamma^*}^{-1}\Gamma$, $\widetilde{B} = -{\Gamma^*}^{-1}B$ and $\widetilde{u}_i = {\Gamma^*}^{-1}u_i$.

In order to ensure that for every admissible value of Γ^*, Γ, B, x and u an unique value for y is associated, the structural form parameters in (6.24) have to fulfill certain restrictions. Such restrictions are called coherency or consistency conditions.

In the following we will demonstrate the coherency problem associated with type II models using a two equation Probit model before generalizing the results to the n equation case.[20] Skipping the index i for convenience and using the indicator function $I(\cdot)$ which equals one if $\cdot$ holds true and zero otherwise, we arrive at

$$y_1^* = \beta_1 x_1 + \gamma_{12} y_2 + u_1 \tag{6.26}$$
$$y_2^* = \beta_2 x_2 + \gamma_{21} y_1 + u_2 \tag{6.27}$$
$$y_i = I(y_i^* \geq 0) \quad i = 1, 2. \tag{6.28}$$

Let us suppose that γ_{12} and γ_{21} are both positive.[21] The correspondence between the latent and observable variables can be depicted using figure 6.3.

Figure 6.3: Simultaneous Probit Model with Multiple Equilibria

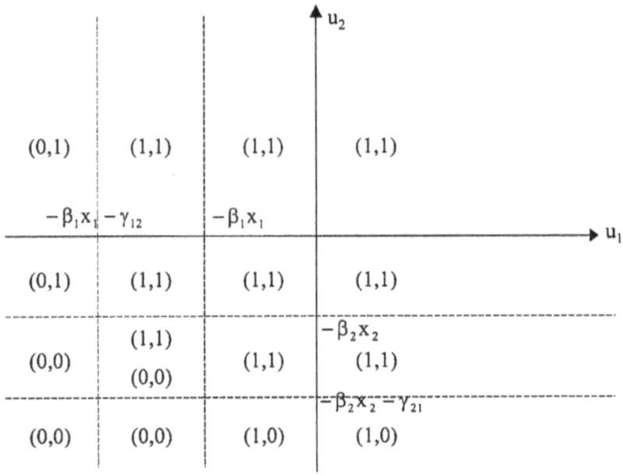

If e.g. u_1 is greater than $-\beta_1 x_1$ then y_1 will equal one regardless of the value of y_2. The same holds if u_2 is greater than $-\beta_2 x_2$ for y_2 so that for all $(u_1, u_2) \in (-\beta_1 x_1, \infty) \times (-\beta_2 x_2, \infty)$ the corresponding value for the observable endogenous variables is $(1, 1)$. In the same manner a unique relationship between (u_1, u_2) and (y_1, y_2) can be derived for the other regions, except for the area $(u_1, u_2) \in (-\beta_1 x_1 - \gamma_{12}, -\beta_1 x_1) \times (-\beta_2 x_2 - \gamma_{21}, -\beta_2 x_2)$.

To show why for this (u_1, u_2) combination multiple equilibria arise, assume that y_2 equals zero. Then y_1 also equals zero since every u_1 in this area is less

[20] See e.g. Butler and Picone (1999).

[21] The same arguments also hold for other signs of γ_{12} and γ_{21}.

than $-\beta_1 x_1$. The same arguments hold for y_2, so that $(0,0)$ is indeed one possible equilibrium.

But now assume that y_2 equals one. Since all u_1 in this area are greater than $-\beta_1 x_1 - \gamma_{12}$, y_1 will equal one, too. The same reasoning can be applied to y_2 so that $(1,1)$ is also an equilibrium.

Therefore two equilibria, $(1,1)$ and $(0,0)$, are consistent with values out of this region. This nonexistence of a unique equilibrium implies that the sum of the probabilities for the distinct outcomes $(0,0)$, $(1,0)$, $(0,1)$ and $(1,1)$ is greater than one therefore making the definition of a random variable impossible.

Now consider the case where e.g. γ_{12} is set equal to zero (see figure 6.4). The correspondence between (u_1, u_2) and (y_1, y_2) now becomes a function, since there are no more regions with multiple or no equilibria. The same holds true if γ_{21} or both γ_{12} and γ_{21} are set to zero. The parameter restriction which ensures that the model (6.26) - (6.28) is coherent therefore can be stated as $\gamma_{12}\gamma_{21} = 0$.

Figure 6.4: Simultaneous Probit Model with Unique Equilibria

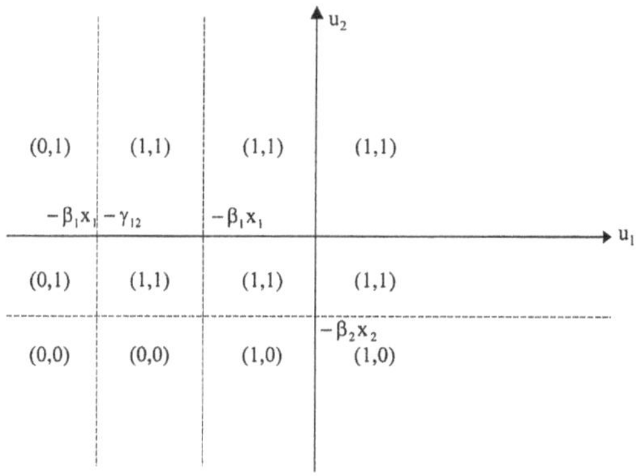

A conclusion from the previous considerations is that a two equation type II Probit model in order to be coherent must feature a recursive structure between the endogenous variables. This restriction often rules out interesting interactions between the variables. In the following we will examine the impacts of this restrictive assumption for our empirical application.

But before, we will extend our findings to the general case. Consider the following n-simultaneous equation type II Probit model:[22]

$$\Gamma^* y^* + \Gamma y + Bx = u \tag{6.29}$$

where all G variables in y are assumed to be truncated and linked via a Probit threshold model with their latent counterpart in y^*. We further assume in (6.29) that only the observable variables appear as explanatory variables by setting $\Gamma^* = I$.

It can be shown that a necessary condition for the coherency of (6.29) is that all principle minors of Γ are equal to zero.[23] In our previously discussed example it holds that:

$$\Gamma = \begin{pmatrix} 0 & -\gamma_{12} \\ -\gamma_{21} & 0 \end{pmatrix}. \tag{6.30}$$

The principle minors of Γ are:

$$|\Gamma_{11}| = 0, \quad |\Gamma_{22}| = 0 \quad \text{and} \quad |\Gamma| = -\gamma_{12}\gamma_{21}. \tag{6.31}$$

The coherency condition therefore is that $\gamma_{12}\gamma_{21}$ must equal zero.

Now assume that only $G^* \leq G$ of the endogenous variables in y are truncated. Partitioning the matrices accordingly, (6.29) can be written as:

$$\begin{pmatrix} y_1^* & y_2 \end{pmatrix} = \begin{pmatrix} y_1 & y_2 \end{pmatrix} \begin{pmatrix} \Gamma_{11} & \Gamma_{12} \\ \Gamma_{21} & \Gamma_{22} \end{pmatrix} + X \begin{pmatrix} B_1 & B_2 \end{pmatrix} + \begin{pmatrix} u_1 & u_2 \end{pmatrix} \tag{6.32}$$

where y_1 are the truncated and $y_2 = y_2^*$ the observable variables.

(6.32) can be written as:

$$y_1^* = y_1\Gamma_{11} + y_2\Gamma_{21} + XB_1 + u_1 \tag{6.33}$$

$$y_2 = y_1\Gamma_{12} + y_2\Gamma_{22} + XB_2 + u_2. \tag{6.34}$$

(6.34) can be analogously written as:

$$y_2 = y_1\Gamma_{12}(I - \Gamma_{22})^{-1} + XB_2(I - \Gamma_{22})^{-1} + u_2(I - \Gamma_{22})^{-1}. \tag{6.35}$$

Inserting (6.35) into (6.33) yields after rearranging:

$$\begin{aligned} y_1^* &= y_1(\Gamma_{11} + \Gamma_{12}(I - \Gamma_{22})^{-1}\Gamma_{21}) + X(B_1 + B_2(I - \Gamma_{22})^{-1}\Gamma_{21}) \\ &\quad + (u_1 + u_2(I - \Gamma_{22})^{-1}\Gamma_{21}). \end{aligned} \tag{6.36}$$

[22] Again we skip the individual index for convenience.

[23] See Schmidt (1990) for a proof.

(6.36) is nothing else than the previously discussed simultaneous Probit type II model with all endogenous variables truncated. Applying the appropriate coherency condition yields that all principle minors of $(\Gamma_{11}+\Gamma_{12}(I-\Gamma_{22})^{-1}\Gamma_{21}$ must be zero.

Until now we have assumed that only the observable variables appear as explanatory variables. Consider now a model where both the latent and the observable variables appear as right hand side variables:

$$\Gamma^* y^* + \Gamma y + Bx = u. \tag{6.37}$$

Solving (6.37) for y^* yields:

$$y^* = \Gamma^{*-1}\Gamma y - \Gamma^{*-1}Bx + \Gamma^{*-1}u \tag{6.38}$$

which is coherent if all principle minors of $\Gamma^{*-1}\Gamma$ are equal to zero. The next table summarizes the coherency conditions for the different models discussed previously.

Table 6.1: Coherency Conditions for Simultaneous Probit Models

Type II Model	Coherency Conditions
Simultaneous Probit Model with All Endogenous Variables Truncated: $Y^* = Y\Gamma + XB + \epsilon.$	All Principle Minors of Γ Must Equal Zero.
Simultaneous Probit Model with Some Endogenous Variables Truncated: $(Y_1^* \;\; Y_2) = (Y_1 \;\; Y_2)\begin{pmatrix}\Gamma_{11} & \Gamma_{12}\\ \Gamma_{21} & \Gamma_{22}\end{pmatrix} + XB + \epsilon.$	All Principle Minors of $I - \Gamma_{11} + \Gamma_{12}(I-\Gamma_{22})^{-1}\Gamma_{21}$ Must Equal Zero.
Simultaneous Probit Model with Both Y and Y^* as Explanatory Variables: $Y^* = Y\Gamma + XB + \epsilon.$	All Principal Minors of $\Gamma(I-\Delta)^{-1}$ Must Equal Zero.

Let us now return to our empirical application as summarized in (6.10) - (6.14). Γ^* and Γ then become:

$$\Gamma^* = \begin{pmatrix} 1 & 0 & 0 & 0\\ 0 & 1 & 0 & 0\\ 0 & 0 & 1 & 0\\ 0 & 0 & 0 & 1 \end{pmatrix} \quad \text{and} \quad \Gamma = \begin{pmatrix} 0 & 0 & \gamma_{13} & \gamma_{14}\\ 0 & 0 & \gamma_{23} & \gamma_{24}\\ 0 & 0 & 0 & \gamma_{34}\\ 0 & 0 & \gamma_{43} & 0 \end{pmatrix}. \tag{6.39}$$

Rearranging Γ according to (6.32), we conclude that all principal minors of:

$$(6.40) \qquad \begin{pmatrix} 0 & \gamma_{43} \\ \gamma_{34} & 0 \end{pmatrix}$$

must be zero, i.e. either $\gamma_{34} = 0$ or $\gamma_{43} = 0$ leading to a recursive structure regarding the innovation decision and introduced organizational changes. In the following we will estimate two different type II models.

The first one (see figure 6.5 left) assumes that the innovation decision might have an impact on introduced organizational changes but not vice versa whereas the second model to be estimated assumes the converse (figure 6.5 right).

Figure 6.5: Type II Dependency Structure and Econometric Models

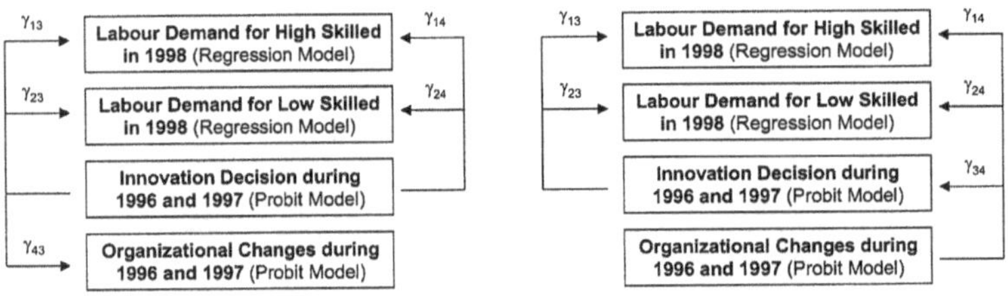

Estimation of Type II Models

For the estimation of the parameters of type II models there are again two different approaches that can be followed. System estimation procedures are computationally more challenging since they require the evaluation of a n-fold integral in the case of a MSEM with n Probit equations. This problem can be avoided by the use of single equation estimation procedures which are in fact instrumental variable estimators. We will describe this approach in the following.[24]

One natural idea for estimating the structural form parameters of the model formulated in (6.10) - (6.13) with the parameter restriction $\gamma_{34}\gamma_{43} = 0$, is to proceed in the following manner: For $\gamma_{43} = 0$, the model can be rewritten as:[25]

(6.41)

$$\begin{aligned} S_{high} &= \gamma_{13}E(INNO) + \gamma_{14}E(ORGA) + \beta'_{high}x_{high} \\ &\quad + \underbrace{(\gamma_{13}(INNO - E(INNO)) + \gamma_{14}(ORGA - E(ORGA)) + u_{high})}_{\equiv v_{high}} \end{aligned}$$

[24]For a thorough discussion on this issue see e.g. Wooldridge (2001), especially chapter 18.

[25]See Wilde (1999).

(6.42)

$$\begin{aligned} S_{low} &= \gamma_{23}E(INNO) + \gamma_{24}E(ORGA) + \beta'_{low}x_{low} \\ &+ \underbrace{(\gamma_{23}(INNO - E(INNO)) + \gamma_{24}(ORGA - E(ORGA)) + u_{low})}_{\equiv v_{low}} \end{aligned}$$

(6.43)

$$INNO^* = \gamma_{34}E(ORGA) + \beta'_{inno}x_{inno} + \underbrace{(\gamma_{34}(ORGA - E(ORGA) + u_{inno}))}_{\equiv v_{inno}}$$

(6.44) $$ORGA^* = \beta'_{orga}x_{orga} + u_{orga}.$$

In a first step equation (6.44) can be estimated using standard ML-methods. The so obtained parameters are then used for estimating equation (6.43) by inserting the following expression for the expected value of $ORGA$

(6.45) $$E(ORGA) = P(ORGA = 1) = P(u_{orga} \geq -\beta'_{orga}x_{orga}) = \Phi(\beta'_{orga}x_{orga})$$

and replacing β_{orga} with the first step estimate $\hat{\beta}_{orga}$.

Finally, the employment share equations in (6.41) and (6.42) can be estimated using (6.45), inserting the following expression for the expected value of $INNO$

(6.46)

$$\begin{aligned} E(INNO) &= P(INNO = 1|ORGA = 1)P(ORGA = 1) \\ &+ P(INNO = 1|ORGA = 0)P(ORGA = 0) \\ &= P(v_{inno} \geq -\gamma_{34} - \beta'_{inno}x_{inno})P(u_{orga} \geq -\beta'_{orga}x_{orga}) \\ &+ P(v_{inno} \geq -\beta'_{inno}x_{inno})P(u_{orga} < -\beta'_{orga}x_{orga}) \\ &= \Phi(\gamma_{34} + \beta'_{inno}x_{inno})\Phi(\beta'_{orga}x_{orga}) \\ &+ \Phi(\beta'_{inno}x_{inno})(1 - \Phi(\beta'_{orga}x_{orga})) \end{aligned}$$

and replacing β_{inno} with $\hat{\beta}_{inno}$ obtained in the previous step.

One problem with this proceeding is, however, that the error terms in (6.41) - (6.43) are not distributed according to a normal distribution since v_{high}, v_{low} and v_{inno} are linear combinations of a continuous normal and a discrete binomial random variable. To account for this misspecification we will estimate the variances of the parameters using the variance estimator proposed by White (1982).[26] The estimation results can be found in the appendix in the column labelled IV-White and will be discussed in section five.

[26] Cf. Wooldridge (2001), p. 625. See also Angrist (1995) and Angrist and Imbens (1995) who discuss under which conditions such a proceeding provides a solution to the selection problem.

6.4 Dataset and First Descriptive Results

We have given an overview of the IAB Establishment Panel already in the previous chapters. The interested reader can find more details about this dataset therein. At this point it is sufficient to note that information about innovations are provided for the years 1993, 1994 and 1998 whereas organizational changes are covered in 1995, 1998 and 2000. Since the 1998 wave contains information about innovation decisions and organizational changes, we will focus our attention especially on this wave.

Regarding the innovation decision the following questions were posed:

- Has your establishment improved an already offered product or service during the last two years? (*Improvement*)

- Has your establishment included a new product or service in its range of products during the last two years that already existed in the market? (*Incremental*)

- Has your establishment included a new product or service in its range of products during the last two years that were new to the market? (*Radical*)

In 1995 and 1998 establishments were also asked if they have introduced one of the following organizational changes during the last two years:

- Reorganization of departments and functional areas (*Reorganization*),

- decentralization of responsibilities and decision-authority (*Decentralization*),

- introduction of group-workplaces with own responsibilities (*Responsibility*)

- introduction of units with own costing and result calculation (*Own Costing*).

The questions about the innovation decisions and organizational changes are retrospective in nature so that we decided to include also information from the 1996 wave and thus to focus our attention on those establishments which participated in 1998 and 1996. The total number of theses establishments amounts to 6,970 with 3,564 coming from West and 3,407 from East Germany.

In the following we will present some first descriptive results for these establishments. Figure E.1 in the appendix shows the distribution of the establishments

Table 6.2: Mean and Standard Deviation for Different Skill Groups of Employees

Employment Group	Mean	Standard Deviation
Low Skilled Blue Collar	60.4	343.57
High Skilled Blue Collar	88.56	443.28
Low Skilled White Collar	19.59	136.05
High Skilled White Collar	173.31	569.64
Apprentices and Working Owners	16.51	51.12
Total Employment	360.72	1,167.24

with regard to the industrial affiliation, whereas figure E.2 shows how they are divided into different size classes. The following table 6.2 shows the average number of employees in the different qualification groups.

Figure E.3 to E.7 in the appendix show the percentage of establishments that introduced different forms of innovations with regard to size class, region and industrial affiliation. It can be concluded that especially large establishments in the primary industry, investments goods industry, finance and insurance are the innovative ones. Additionally, West German establishments compared to their eastern counterparts are more innovative.

Estimating the Pearson correlation coefficient between the different forms of innovations reveals that there are positive and significant correlations between them, especially between an improvement of existing products and services and an incremental product innovation (see table E.16 in the appendix). Due to these high positive correlations we decided to combine the different innovation forms to one single dummy variable, $INNO$, which equals one if at least one of the innovation forms is realized and zero otherwise.

In the following we will examine organizational changes introduced by the establishments. Figures E.8 to E.13 in the appendix show the percentage of establishments that introduced different organizational changes with regard to size class, region and industrial affiliation. Again it is striking that especially large establishments in the electricity/mining segment, primary industry, investments goods industry and finance segment are those that introduced organizational changes. Additionally, West German establishments are more likely to introduce organizational changes compared to their eastern counterparts.

Estimating the Pearson correlation coefficient between the different organizational changes reveals that there are positive and significant correlations between them as it was the case for the different forms of innovations (see table E.17 in the appendix). Due to these high positive correlations we decided to combine the different organizational changes to one single dummy variable, $ORGA$, which equals one if at least one of the changes is realized and zero otherwise.

Having provided a first overview about the establishments which have introduced innovations and organizational changes, finally we will take a first look at the employment effects. To this aim we compare the employment shares for high and low skilled employees in 1998 for those establishments which have introduced innovations and organizational changes with establishments which have not.

More specifically we will distinguish the following establishments: Establishments which have introduced product innovations and organizational changes ($INNO = 1, ORGA = 1$), establishments which introduced either innovations or organizational changes ($INNO = 1, ORGA = 0$ and $INNO = 0, ORGA = 1$) and establishments which have introduced neither innovations nor organizational changes ($INNO = 0, ORGA = 0$). Table E.18 in the appendix contains the percentages of the total 6,970 establishments falling in each category. The Pearson correlation coefficient between $INNO$ and $ORGA$ equals 0.39 and is highly significant (significance level < 0.001) indicating that innovations and organizational changes are indeed, as already stated above, closely related to each other.[27]

Table 6.4 contains the means and standard deviations for the different employment shares.[28] One striking feature is that establishments which introduced either innovations or organizational changes are the ones with the higher share of skilled employees in contrast to establishments which have introduced neither innovations nor organizational changes. This impression is confirmed by a look at the development of the shares for high and low skilled employees between 1998 and 1996 (see table 6.4). Again establishments which introduced either innovations or organizational changes are the ones with a larger increase of high skilled employees and a larger decrease of low skilled employees.

---

[27]One might suspect that technological and organizational changes are influenced by one common factor and thus can not be separated anyway. Conducting a Factor analysis yields, however, that the correlation matrix for the innovation and organizational change decision is not eligible for a Factor analysis. E.g. the Kaiser-Meyer-Olkin measure of sampling adequacy is in our case 0.5 and therefore too small for a satisfactory analysis. Analysing the anti-image-matrix and the inverse of the correlation matrix yields similar results. We therefore decided to maintain our model.

[28]Note that the shares do not sum to 100% since we have omitted other employment groups such as apprentices and working owners.

Table 6.3: Development of Different Employment Groups (Standard Deviations in Brackets)

Employment Group	$INNO = 0$ $ORGA = 0$	$INNO = 1$ $ORGA = 0$	$INNO = 0$ $ORGA = 1$	$INNO = 1$ $ORGA = 1$
Share High Skilled	68% (32.66)	69% (29.21)	74% (29.36)	73% (24.99)
Growth High Skilled	0.78% (32.52)	1.74% (26.64)	2.39% (24.07)	1.22% (19.73)
Share Low Skilled	26% (32.90)	24% (29.17)	20% (26.59)	22% (24.97)
Growth Low Skilled	-1.30% (31.75)	-2.09% (26.78)	-2.49% (23.23)	-1.90% (19.15)

These descriptive results give a first evidence for the hypothesis that there is a skill biased technological and organizational change in favor of high skilled employees. In the following we will test whether these first descriptive results are confirmed by the estimation results of the econometric models.

6.5 Estimation Results

In this chapter we were mainly interested in the employment effects of innovations and organizational changes. To this aim several econometric models were estimated using different estimation methods, ranging from separate estimation of the employment share, innovation and organizational change equation to models and methods which take the endogeneity of innovations and organizational changes into account.

Thereby we employed two different types of models, namely type I models which assume that it is the propensity to innovate or to introduce organizational changes which might have an impact on the demand for heterogenous labor, and type II models which on the other hand assume that it is the actual decision that has an effect. The detailed estimation results can be found in the appendix (tables E.4 to E.15). In the following we will summarize the main findings and derive some stylized facts. We start with the relations between the endogenous variables, namely share of high and low skilled employees, innovations and organizational changes (see figure 6.6).

Figure 6.6: Stylized Estimation Results for the Different Models

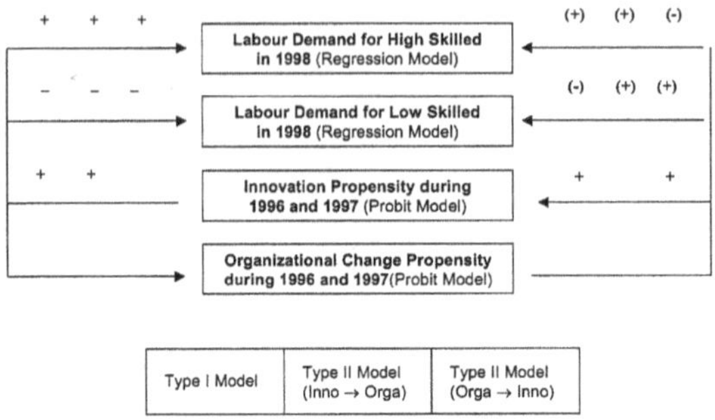

Note: Only the Signs are Reported, Insignificant Parameters in Brackets

By comparing the estimation results for the different models, the following stylized facts can be derived: Both type I and type II models support the hypothesis about a skill biased technological change in favor of high skilled employees. For the type I model, estimation based on the procedure proposed by Amemiya yields no significant results whereas the estimation based on Nelson-Olson confirms the OLS estimation results, namely that innovations have a positive impact on high skilled and a negative on low skilled employees. The same holds for both recursive type II models that were estimated, the first which assumes that innovations have an impact on organizational changes ($INNO \rightarrow ORGA$) and the second which assumes the opposite, i.e. $ORGA \rightarrow INNO$.

At this point one natural arising question is the following: Do only implemented innovations and organizational changes have an impact on the labor demand or even the propensity to do so, i.e. are type I or type II models more appropriate? To this aim we will conduct a simple test proposed by Blundell and Smith (1993) and Davidson and MacKinnon (1981) that will allow us to discriminate between them.

At first we will test whether the type I formulation of our model is appropriate, i.e. our null hypothesis is:

$$H_0: \quad \text{Type I model is the true model} \tag{6.47}$$

and the two alternative hypotheses are:

(6.48) H_1 : Type II model ($INNO \rightarrow ORGA$) is the true model

(6.49) H_2 : Type II model ($ORGA \rightarrow INNO$) is the true model.

If the type I formulation is the true model, then for the share equation for high and low skilled employees, S_{high} and S_{low}, explicit reduced form equations exist:

$$\begin{aligned} S_{high} &= \pi'_{high}x + v_{high} \\ S_{low} &= \pi'_{low}x + v_{low}. \end{aligned} \tag{6.50}$$

If, however, the type II formulation is the correct one, then only a semireduced form is available which also contains the endogenous dummy variables:

$$\begin{aligned} S_{high} &= \pi'_{high}x + \gamma_{13}INNO + \gamma_{14}ORGA + v_{high} \\ S_{low} &= \pi'_{low}x + \gamma_{23}INNO + \gamma_{24}ORGA + v_{low}. \end{aligned} \tag{6.51}$$

One obvious way to test H_0 therefore consists in estimating (6.51) by replacing $INNO$ and $ORGA$ by the appropriate estimated predictions under the type II formulation and test if $\gamma_{13} = \gamma_{14} = 0$ and $\gamma_{23} = \gamma_{24} = 0$, respectively.[29]

The results can be found in the following table 6.4. They indicate that H_0 can not be rejected, i.e. the type I formulation is appropriate for the share equation of high and low skilled employees. We conclude that it is not only the actual decision to innovate and to introduce organizational changes that affects the labor demand but also the propensity to do so.

Table 6.4: Model Specification Test: H_0 : Type I Model is the True Model

$H_{alternative}$	F-Value	p-Value
Type II ($INNO \rightarrow ORGA$) is the True Model for High Skilled	0.56	0.57
Type II ($INNO \rightarrow ORGA$) is the True Model for Low Skilled	0.96	0.38
Type II ($ORGA \rightarrow INNO$) is the True Model for High Skilled	1.64	0.19
Type II ($ORGA \rightarrow INNO$) is the True Model for Low Skilled	0.70	0.50

Next we will test if our data are also consistent with the type II formulation. The type II formulation can be tested by estimating:[30]

$$\begin{aligned} S_{high} &= \gamma_{13}INNO + \gamma_{14}ORGA + \beta'_{high}x_{high} + \delta_{high}\hat{S}_{high} + u_{high} \\ S_{low} &= \gamma_{23}INNO + \gamma_{24}ORGA + \beta'_{low}x_{low} + \delta_{low}\hat{S}_{low} + u_{low} \end{aligned} \tag{6.52}$$

[29]For the type II model e.g. with $\gamma_{43} = 0$, i.e. the $INNO \rightarrow ORGA$ model, $E(ORGA) = \Phi(\hat{\beta}'_{orga}x_{orga})$ and $E(INNO) = \Phi(\hat{\gamma}_{34} + \hat{\beta}'_{inno}x_{inno})\Phi(\hat{\beta}'_{orga}x_{orga}) + \Phi(\hat{\beta}'_{inno}x_{inno})(1 - \Phi(\hat{\beta}'_{orga}x_{orga})$.

[30]See Blundell and Smith (1993).

and testing if $\delta_{high} = 0$ and $\delta_{low} = 0$ respectively. Thereby $\hat{S}_{high}$ and $\hat{S}_{low}$ are the estimated predictions under the type I explicit linear reduced form for S_{high} and S_{low}. The following table contains the t-values and p-values for δ_{high} and δ_{low}.

Table 6.5: Model Specification Test: H_0 : Type II Model is the True Model

$H_{alternative}$	t-Value	p-Value
Type I is the True Model for High Skilled	34.67	<0.0001
Type I is the True Model for Low Skilled	32.66	<0.0001

Since the null hypothesis is rejected on a high significance level we conclude that type II models are not appropriate in our context compared with type I models. It seems therefore that the censoring of the $INNO$ and $ORGA$ variables acts as an observation constraint for the econometrician but not for the establishment. Establishments seem to have the choice between a whole scope of innovation and organizational change decisions which affect the labor demand and not only between the ones observed in the dataset.

We continue with other factors that have a significant impact on the labor demand. These are first of all investments in information and communication technology. Again the effect is diametral regarding different skill groups of employees, i.e. positive for high and negative for low skilled. The state of the technology of an establishment works in the same direction: The newer the technology, the more likely high skilled employees will be employed.

With regard to organizational changes no significant effects on the labor demand can be observed, independent of the econometric model. We thus found no empirical support for our initially formulated hypothesis that organizational changes might also have a skill biased impact on the labor demand.[31]

[31] Our results are in conflict with previous empirical findings, like e.g. Caroli and Reenen (2001). They have analyzed British and French establishments and showed that organizational changes and skills are complements. For the United States Bresnahan, Brynjolfsson, and Hitt (2002) have also found a complementary relationship between investments in ICT, organizational changes and high skilled employees. Similar results for Germany are obtained by Falk (2001) who found that organizational changes are mainly driven by ICT and training expenditures and have a positive impact on expected employment for all skill groups except for unskilled labor. The results also stand in contrast to the results reported in the next chapter which makes use of a linked employer employee dataset. This dataset contains detailed information about the formal education of employees. One explanation for the differing results therefore might be the crude qualification measures used in the IAB Establishment Panel.

Another striking feature of our estimation results concerns the relation between innovations and organizational changes. Any estimated model points out that introduced innovations and organizational changes are closely related.

The major factors driving innovations are the age of the technology used (a modern equipment increases the probability to innovate), the existence of a research & development and a market research department, the competition pressure (with an increasing competition the probability to innovate increases) and the size of the establishment (larger establishments are more likely to innovate than smaller ones). Interestingly the portion of high skilled employees has no effect on the innovation activities.

Regarding the decision of establishments to introduce organizational changes the following factors can be identified: Investments in information and communication technology increase the probability to introduce organizational changes thereby supporting our initially formulated hypothesis about the relation between technological and organizational change. Other driving factors are business development (an insufficient development increases the probability to introduce organizational changes), the existence of a workers and staff council and the size of the establishment which both increase the probability to introduce organizational changes. Again the portion of skilled employees has no significant impact.

6.6 Summary and Conclusions

The purpose of this chapter was to answer the following questions: What is the effect of technological change on the labor demand, especially for different skill groups of employees? Does the numerous recently introduced organizational changes also have an effect on the labor demand? Which effect prevails: Innovations or organizational changes? If technological progress and organizational changes are indeed skill biased as often stated, what is the net effect on the total labor demand?

Another interesting point is the following: Do only implemented innovations and organizational changes have an impact on labor demand or even the propensity to do so, i.e. are type I or type II models more appropriate?

Finally, we wanted to know if certain organizational environments are more likely to generate innovations or if innovations entail organizational changes?

Our results indicate that product innovations operate in an asymmetrical way, increasing the labor demand for high and reducing it for low skilled employees.

For organizational changes we did not find a similar effect. By comparing the absolute numbers of the innovation effect on the labor demand for high and low skilled employees, it seems that the total labor demand effect is almost zero, but noticeable larger for type II models than for the type I model. This means that implemented innovations have a more severe impact on labor demand than only the propensity to innovate.

By conducting several specification tests in order to discriminate between type I and type II models, we arrived at the insight that models, which assume that the propensity to innovate and to introduce organizational changes have an impact on the labor demand, are more appropriate than models which assume that it is the actual decision to do so. It thus seems that the censoring of the variables acts as an observational constraint for the econometrician but not for the set of choices of the individual establishment.

One additional finding is the close relation between innovations and organizational changes. Every estimated model pointed out that innovations are always associated with organizational changes and vice versa. In the following chapter we will analyze if these results are confirmed using a more detailed linked employer employee dataset.

Chapter 7

Digging into Deep: Results Using a Linked Employer Employee Dataset

7.1 Introduction

In this chapter we are still concerned with the analysis of labor demand and especially with the impact of technological and organizational changes on high and low skilled employees. In the previous chapter we have estimated various mixed simultaneous equation models in order to control for a possible sample selection process. The estimation results indicated that product innovations and organizational changes are skill biased in favor of high skilled employees, albeit organizational changes only on a low significance level.

These empirical findings, however, have to be treated with care. The first point concerns the crude measure for different skills. The IAB Establishment Panel does not contain any information about the educational attainment or the actual activities of the employees. Furthermore due to data limitations we were forced to approximate wages for different skill groups with industry dummies. We will accommodate this shortcoming in the following using an unique linked employer employee dataset for Germany which contains detailed information about wages and formal educational attainments of employees.[1]

The availability of such datasets has increased immensely in the last couple of years. A major advantage of this kind of datasets is the fact that interactions between individuals and firms can be modelled explicitly and therefore an analysis

[1]This chapter is based on Bellmann, Caliendo, Hujer, and Radić (2002).

from the labor supply as well as from the labor demand side is possible.[2]

For Germany a linked employer employee dataset is available through the combination of the IAB Establishment Panel which contains detailed information about the employer side and the Employment Statistics Register covering the employees. We have already used this dataset in chapter five where we analyzed the process of organizational changes.

The aim of this chapter is to make another contribution to the question on the employment effects of technological and organizational changes using such a rich linked employer employee dataset. We proceed as follows: In the next section the theoretical basis and the empirical implementation is provided. Section three gives a more detailed description of the dataset and first descriptive results. In chapter five the econometric estimation results are commented and, finally, chapter six summarizes.

7.2 Theoretical Considerations and Econometric Modelling

In the following, we will present some theoretical thoughts on the explanation of the demand for various production factors which will serve as a basis for the subsequent empirical implementation. These considerations are similar to the proceeding of the previous chapter and hence fail briefly.

Labor Demand

Given that certain regularity conditions and thus the duality theorem hold, the production behavior of a firm can be equivalently described by its production or cost function.[3] Every information contained in the production function is also contained in the cost function and vice versa. The choice of the approach can then be made on basis of the available data. Since we are using a cross section dataset, it is reasonable to assume that for a given year the prices for output and input factors are exogenously given and hence in the following we will estimate a cost function.[4]

[2]Abowd and Kramarz (1999a,199b) or Haltiwanger et. al. (1999) give a detailed overview of different linked datasets, specific questions, statistical modelling and potential interpretations. Moreover, they summarize first results from international studies.

[3]For a formal proof of the duality theorem see for example Varian (1984) or Diewert (1971).

[4]for more details see e.g. Berndt (1991).

A number of cost function have been suggested in the literature.[5] In the previous chapter we have used the translog function. Another very flexible one is the generalized Leontief function introduced by Diewert (1971,1987). In contrast to the Cobb-Douglas or CES function the elasticities between different input factors are not restricted a priori to certain values. Besides that, additional explanatory factors can be simply included as linear terms. Three variable input factors are assumed: Low skilled, medium skilled and high skilled employees. Low skilled include employees without formal education, medium skilled are employees with vocational training whereas employees with a university or equivalent degree constitute the group of high skilled employees.

Additionally, we will consider a set of quasi-fixed input factors, namely capital, product innovations and organizational changes.[6] Capital stock is approximated by the sum of investments in the previous year. The operationalization of the other factors will be explained in more detail in the following section. The K exogenous variables can be collected into a vector Z. With w_H, w_M and w_L as the wage rates for high, medium and low skilled employees, the variable costs of a representative firm C conditional on a production level Y can then be written as:[7]

$$(7.1) \qquad \begin{aligned} C &= C(w_H, w_M, w_L, Z, Y) \\ &= h(Y)\left[\sum_{i=1}^{3}\sum_{j=1}^{3}\alpha_{ij}(w_i w_j)^{\frac{1}{2}}\right] + h(Y)\left[\sum_{i=1}^{3}\sum_{j=1}^{K}\beta_{ij} w_i Z_j\right], \end{aligned}$$

for $i = H, M$ and L. We assume that the following symmetry conditions are met: $\alpha_{ij} = \alpha_{ji}, \forall i, j$. $h(Y)$ is a monotonic function of Y with $h(0) = 0$ and $h(Y) \to \infty$ for $Y \to \infty$. Typically, it is assumed that $h(Y) = Y$ thereby implying constant returns to scale. According to Shephard's lemma the first derivative of the cost function with respect to w_i is equal to the optimal demand for the i-th production factor, X_i, given a certain output:

$$(7.2) \qquad \frac{\partial C}{\partial w_i} = X_i = Y\left[\sum_{i=1}^{3}\alpha_{ij}\left(\frac{w_j}{w_i}\right)^{\frac{1}{2}}\right] + Y\left[\sum_{j=1}^{K}\beta_{ij} Z_j\right], \text{ for } i = H, M, L.$$

[5]For an overview see e.g. Hamermesh (1986,1993).

[6]We have also controlled for industrial affiliation in every estimated equation.

[7]Production level was measured by the actual business volume in million DM. Since except of labor no other variables production functions were considered, using business volume instead of value added poses no problem.

Division of the above factor demand functions by Y yields equivalent functions which, however, can be easier estimated since they are linear in the parameters:

$$(7.3)\qquad \frac{X_i}{Y} = \left[\sum_{i=1}^{3} \alpha_{ij} \left(\frac{w_j}{w_i}\right)^{\frac{1}{2}}\right] + \left[\sum_{j=1}^{K} \beta_{ij} Z_j\right], \text{ for } i = H, M, L.$$

Once the parameters of the factor demand functions in 7.3 are estimated by means of adequate econometric methods, various elasticities between the production factors can be derived. The Hicks-Allen elasticity e.g. indicates, how the input ratio of two production factors changes if their relative prices change given constant output. It therefore indicates how easy a firm can react to relative price changes if the only adjustment mechanism is the quantity of the two production factors. In general, the Hicks-Allen cross elasticity is defined as follows:

$$(7.4)\qquad \sigma_{ij} = \frac{\partial lnX_i/lnX_j}{\partial lnP_j/lnP_i} = \frac{C \cdot C_{ij}}{C_i \cdot C_j}, \text{ for } i \neq j$$

where C_{ij} is the cross derivative of the cost function with respect to the i-th and j-th production factor price. σ_{ij} equals one if the production factors are perfect substitutes. For the generalized Leontief cost function we obtain:

$$(7.5)\qquad \sigma_{ij} = \begin{cases} \dfrac{\frac{1}{2}C\alpha_{ij}(w_i w_j)^{-\frac{1}{2}}}{YX_iX_j}, & \text{for } i \neq j \\ \dfrac{-\frac{1}{2}C \sum\limits_{j=1, j\neq i} \alpha_{ij} w_j^{\frac{1}{2}} w_i^{-\frac{3}{2}}}{YX_iX_j}, & \text{for } i = j. \end{cases}$$

Multiplication of the Hicks-Allen elasticities with the cost share of the other production factor, S_j, yields the own price and cross price elasticities:

$$(7.6)\qquad \epsilon_{ij} = \frac{\partial lnX_i/X_i}{\partial lnP_j/P_j} = \sigma_{ij} S_j.$$

Substituting the j-th factor with the i-th factor is more difficult the lower the cost share S_j of the j-th product. In this case a price decrease of the j-th production factor would have only a minor effect on the already abundantly used production factor i. As a consequence the elasticity ϵ_{ij} decreases. Using the generalized Leontief cost function yields:

$$(7.7)\qquad \epsilon_{ij} = \begin{cases} \dfrac{\frac{1}{2}\alpha_{ij}\left(\frac{w_i}{w_j}\right)^{-\frac{1}{2}}}{X_i}, & \text{for } i \neq j \\ \dfrac{-\frac{1}{2}\sum\limits_{j=1, j\neq i} \alpha_{ij}\left(\frac{w_i}{w_j}\right)^{-\frac{1}{2}}}{X_i}, & \text{for } i = j. \end{cases}$$

Two production factors i and j are said to be p-substitutes if the relative increase in the price of one factor leads to an increase in demand for the other, i.e. $\epsilon_{ij} > 0$. If on the other hand $\epsilon_{ij} < 0$ holds, the production factors are referred to as p-complements.[8] The estimated parameters for the quasi-fixed exogenous factors directly reflect the impact on the input-output-coefficients for the different qualification groups.

An obvious empirical strategy consists of estimating the above factor demand system with adequate econometric methods like e.g. Zellner's SURE method (cf. Zellner (1962)). A potential problem of this approach, however, is the potential endogeneity of the firm's decision to innovate or to implement organizational changes. Although these variables refer in our empirical implementation to decisions which have been made in the last two years, the presumption remains that they are still correlated with the factor demand equations. Therefore, the two decisions will be modelled explicitly in a simultaneous equation framework.

Innovation Decision

Regarding the innovation decision, our dataset contains the same information on product innovations for the years 1996 and 1997 like the one we have used in the previous chapter. In detail, the firms were asked whether they had improved a product or service or brought a new product or service on the market (new to the company or even to the market). Due to positive correlations between the different types of innovations, a dummy variable $INNO$ was generated, which takes on the value one if at least one of these three types of innovations was introduced and zero otherwise. Analogously to the proceeding in the previous chapter, the innovation decision is again modelled using a Probit model:

$$INNO = \begin{cases} 1, \text{ if } INNO^* = \beta_I' x_I + \gamma_{I0} ORGA + \epsilon_I \geq 0 \\ 0, \text{ otherwise.} \end{cases} \tag{7.8}$$

The latent propensity to innovate, $INNO^*$, thereby depends on a set of exogenous establishment and market specific factors captured in the vector x_I. We used the same internal and external factors like in the previous chapter, i.e. qualification structure of the employees, the existence of a R&D department, research cooperations with other institutions, degree of competition or profitability of a firm.[9]

[8] Cf. Hamermesh (1991).

[9] See also table E.2 in the appendix to chapter six.

Again, special attention is paid to organizational changes, $ORGA$, especially to the question whether flexible and less hierarchical organizational structure increase the innovation propensity of a firm or not.

Organizational Changes

Organizational changes can be regarded as disembodied process innovations.[10] Their impact on company performance and labor demand has become more important in the last couple of years which can be explained by the fact that in times of rapid economic changes companies with more flexible organizational structures are more competitive. As a result more and more companies changed their organization and gave up the classic, centralized so-called 'Tayloristic' organization, characterized by a strict definition of functions and a specialization of tasks.[11] Instead a tendency towards more 'Holistic' organizational structures can be observed which are characterized by the following features:

- Flatter hierarchies
- Decentralization of responsibility and decision authority
- Introduction of team work and units with more responsibility and own costing

The dataset contains information on such introduced organizational changes during the years 1996 and 1997.[12] Due to a high degree of correlation these changes are collected in a dummy variable $ORGA$ which equals one if the latent propensity for organizational changes exceeds a certain threshold value:

$$ORGA = \begin{cases} 1, \text{ if } ORGA^* = \beta'_O x_O + \gamma_{OI} INNO + \epsilon_O \geq 0 \\ 0, \text{ otherwise.} \end{cases} \tag{7.9}$$

Econometric Model

Collecting the labor demand equations for high, medium and low skilled employees and combining them with the equations for the innovation and organizational

[10] See also chapter one of this thesis on the classification of innovations.

[11] See also chapter five of this thesis.

[12] More details can be found in the previous chapter. See especially table E.3 for more information about the included exogenous variables.

change decision, yields the following econometric model:[13]

(7.10)

$$\begin{aligned}
\frac{X_H}{Y} &= \alpha_{HH} + \alpha_{HM}\left(\frac{w_M}{w_H}\right)^{\frac{1}{2}} + \alpha_{HL}\left(\frac{w_L}{w_H}\right)^{\frac{1}{2}} \\
&\quad + \beta_{HK}K + \beta_{HI}INNO^* + \beta_{HO}ORGA^* + \epsilon_H \\
\frac{X_M}{Y} &= \alpha_{MM} + \alpha_{MH}\left(\frac{w_H}{w_M}\right)^{\frac{1}{2}} + \alpha_{ML}\left(\frac{w_L}{w_M}\right)^{\frac{1}{2}} \\
&\quad + \beta_{MK}K + \beta_{MI}INNO^* + \beta_{MO}ORGA^* + \epsilon_M \\
\frac{X_L}{Y} &= \alpha_{LL} + \alpha_{LH}\left(\frac{w_H}{w_L}\right)^{\frac{1}{2}} + \alpha_{LM}\left(\frac{w_M}{w_L}\right)^{\frac{1}{2}} \\
&\quad + \beta_{LK}K + \beta_{LI}INNO^* + \beta_{LO}ORGA^* + \epsilon_L \\
INNO &= \begin{cases} 1, \text{ if } INNO^* = \beta_I' x_I + \gamma_{IO} ORGA^* + \epsilon_I \geq 0 \\ 0, \text{ otherwise} \end{cases} \\
ORGA &= \begin{cases} 1, \text{ if } ORGA^* = \beta_O' x_O + \gamma_{OI} INNO^* + \epsilon_O \geq 0 \\ 0, \text{ otherwise.} \end{cases}
\end{aligned}$$

The above model constitutes a type I mixed simultaneous equation model. In the previous chapter we have estimated such a type I and also a type II model where instead of the latent variables $INNO^*$ and $ORGA^*$ their observable counterparts $INNO$ and $ORGA$ have been included. One problem with type II models are the additional consistency conditions which they require. In our case these conditions amount to impose a recursive relation between the innovation and the organizational change decision. Since we were able to reject the type II formulation in favor of the type I model in the previous chapter, in the following we will only estimate the type I model formulation as given in (7.10).

The estimation consists of the following steps: First, according to the Nelson-Olson procedure, we estimated the reduced form parameters of (7.8) and (7.9) in order to obtain predicted values for $INNO^*$ and $ORGA^*$. Plugging these predicted values in the labor demand equations the structural form parameters were obtained using Zellner's SURE. The so obtained estimation results can be used to calculate the various elasticities which were introduced previously. Since these elasticities are highly nonlinear functions of the estimated parameters their standard errors were bootstrapped.

[13]Note that since we are not working with employment shares, the labor demand equations are linearly independent.

7.3 Dataset and First Descriptive Results

We will estimate the econometric model given in (7.10) using an unique linked employer employee dataset for Germany which was already used in chapter five of this thesis. Information about the employer side is provided by the IAB Panel which contains information about innovative behavior (for the years 1993, 1994 and 1998) and organizational changes (for the years 1995, 1998, 2000). Additionally, it offers us other control variables, e.g. information about business and investment policy. Since this dataset has been discussed in previous chapters we will not go into more detail at this point.

The second data source is the Employment Statistics Register which is an administrative dataset containing individual information on wages of all employees which are subject to the compulsory social security system.[14] Both datasets can be merged using an identical establishment number yielding a linked employer employee dataset for Germany which contains detailed information on wage levels and employment shares for different qualification groups.

One peculiarity of the employee dataset is that in 1999 the reporting procedure changed. This change of the Data Recording and Transmission Decree (Datenerfassungs- und Datenübermittlungsverordnung, 'DEÜV') influenced the information in the Employment Statistics Register for the year 1998 since it is possible that during the transition period establishments either reported according to the old or the new reporting procedure thus leading to multiple employment entries. We therefore included only those annual reports that were non-ambiguous.[15]

Information on innovations as well as organizational changes are provided for the year 1998 but refer to the two previous years 1996 and 1997. In a first step we therefore merged establishment and individual data for those establishments which participated in 1996 and 1998 at the IAB Establishment Panel. Information from the 1996 wave were used to construct the exogenous variables. We arrive at a number of 5,600 establishments for which we calculated and merged the aggregated average daily wages and employment shares for three qualification

[14]For more details on this dataset see chapter five of this thesis and the literature cited therein.

[15]Although the absolute number of observations reduces considerable by round 20%, the employment shares regarding different qualification structures remained stable compared to previous years. Another point regarding the Employment Statistics Register concerns the censoring of wages. Wages which are higher than the limit up to which social security payments are compulsory (8,400 and 7,000 DM in 1998 for West and East Germany, respectively) are not included in the dataset. For more details see Neidert (1998).

groups: High, medium and low skilled employees. As has been mentioned before, employees have been categorized as high skilled if they had obtained a university degree or equivalent, medium skilled if they had completed vocational training and low skilled if they had no formal education at all.

In the following, descriptive results for the year 1998 are presented separately for East and West German establishments. Table 7.1 e.g. shows that the share of high skilled employees in East Germany is higher but that the average number of total employees is lower.

Table 7.1: Average Total Employment and Employment Shares

	West Germany		East Germany	
Employment Group	Mean	SE	Mean	SE
High Skilled	39.67	148.79	23.11	81.76
Share	7.06%	13.36%	11.98%	19.89%
Medium Skilled	246.64	699.36	96.7	232.39
Share	66.15%	26.03%	69.94%	29.04%
Low Skilled	85.32	249.22	19.18	68.29
Share	26.79%	25.79%	18.07%	26.59%
Total Sum	371.63	1,024.08	138.99	325.73

Figure 7.1 gives an idea about the development of the different qualification shares in West Germany for the period from 1993 to 1998.[16] Even within this short period of time a trend towards higher qualification in the establishments can be recognized. The share of high skilled employees increased by 1.93 percentage points during this period from 9.51% to 11.44% while in the same time the share of low skilled employees decreased dramatically by nearly 5 percentage points.

In the following we will conduct a shift-share-analysis where the total increase of high skilled employees, ΔH, is decomposed into one part that was observed within the industrial sectors, the so called 'within'-component ($\sum_i \overline{w}_i \Delta H_i$), and a second component that can be attributed to shifts between various industrial sectors, the so called 'between'-component ($\sum_i \Delta w_i \overline{H}_i$). Like already discussed in chapter one, such a shift-share-analysis can be used to test the international trade hypothesis of increasing labor demand.

[16]East German establishments were only included since 1996 in the dataset. Note that these figures differ from the one in table 7.1 since they are based on establishments which have participated both in 1996 and 1998.

Figure 7.1: Development of the Shares of High and Low Skilled Employees in West Germany

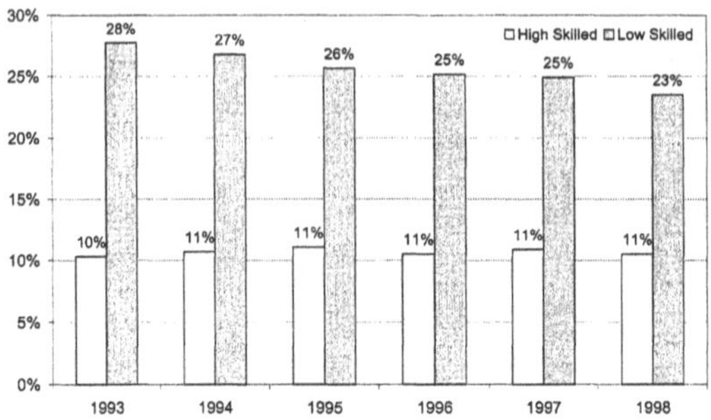

The decomposition is conducted using the following formula:[17]

$$\Delta H = \sum_i \overline{w}_i \Delta H_i + \sum_i \Delta w_i \overline{H}_i \tag{7.11}$$

where

H = Share of high skilled employees
w_i = Share of employees in the i-th industrial sector
H_i = Share of high skilled employees in the i-th industrial sector
$\overline{H}_i$ = Average of the share of high skilled employees over time
$\overline{w}_i$ = Average of the employment share over time

The figures in table 7.2 make clear that the upgrading of the workforce is clearly a within industry effect. For the group of high skilled employees 91.88% of the total increase of 1.77% is due to a within increase while for the group of high and medium skilled the within component amounts to 85.99%. These figures exclude the globalization hypothesis as a potential explanation for an increasing labor demand for high skilled employees. If namely an increasing international trade with low wage countries was responsible for a deterioration of the labor market situation of low skilled employees, the phenomenon of higher qualification

[17]For more details see e.g. Berman, Bound, and Griliches (1994). We have considered the following industries: Agriculture/forestry, electricity/mining, primary industry, investments goods industry, consumer goods industry, construction, transportation, financing, insurance, consumer services, education, health services, other services, non profit organizations, government and administration.

would be attributable primarily to movements from less to more skill intensive sectors.

Table 7.2: Results of the Shift-Share-Analysis

Employment Shares	1993	1998	Total	Within	Between	Within (%)
High Skilled	9.51%	11.44%	1.93%	1.77%	0.16%	91.88%
High and Medium Skilled	72.73%	76.67%	3.94%	3.39%	0.55%	85.99%

In order to get a first impression about the employment effects of innovations and organizational changes, in the following we will take a look at the shares of different qualification groups for different subgroups of establishments (see table 7.3).[18] We differentiate between establishments which have introduced neither product innovations nor organizational changes and establishments which have introduced either one or both. One can see clearly that establishments which have introduced innovations or organizational changes tend to have also a larger share of high skilled employees. This seems to give a hint to the existence of a skill biased technological and organizational change. The next chapter investigates whether these descriptive results can be affirmed by the econometric model.

Table 7.3: Share for Different Skill Groups (Standard Deviations in Parentheses)

Employment Group	All	$INNO = 0$ $ORGA = 0$	$INNO = 1$ $ORGA = 0$	$INNO = 0$ $ORGA = 1$	$INNO = 1$ $ORGA = 1$
High Skilled	9.57%	7.76%	8.52%	10.44%	11.34%
	(17.18%)	(17.82%)	(17.03%)	(16.06%)	(16.43%)
Medium-Skilled	68.09%	69.53%	66.31%	69.70%	66.41%
	(27.67%)	(31.26%)	(29.47%)	(24.62%)	(23.75%)
Low-Skilled	22.34%	22.72%	25.17%	19.86%	22.25%
	(26.56%)	(30.08%)	(28.55%)	(23.43%)	(22.84%)

[18]Differences to the previous figures are due to the fact that we had to include also information about the innovation and organizational change decisions of establishments. The resulting missing values are responsible for the small deviations.

7.4 Estimation Results

As opposed to the descriptive results, for the econometric estimation we omitted establishments from the agricultural and public sector since it can be assumed that the innovative and organizational behavior in these groups differs distinctly from other industrial sectors. Moreover, only those establishments have been considered which have at least two persons in each qualification group.[19] Detailed results can be found in the appendix (tables F.1 to F.6), whereas Figure 5 contains the most important stylized facts.

Figure 7.2: Stylized Estimation Results

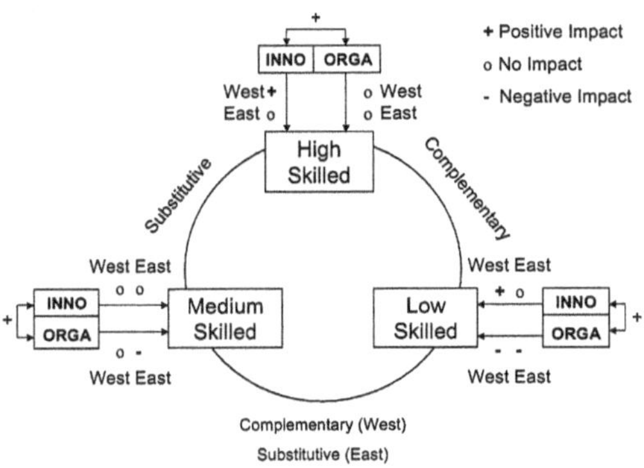

A first result is the strong relationship between innovations and organizational changes. Organizational changes like flatter hierarchies or the introduction of flexible workplaces with more responsibility promote product innovations. On the other hand product innovations require holistic organizational structures.[20] Only in case of establishments from West Germany product innovations seem to have no significant influence on organizational changes. Other factors influencing both decisions are the state of technology, degree of competition as well as investments into information and communication technology, which all have a positive impact on both decisions.

[19]For a similar proceeding see e.g. Bellmann and Schank (2002). A rational for this exclusion is the fact that small establishments with only a few employees have no substitution possibilities at all.

[20]This finding is also confirmed by other studies. See e.g. Wolf and Zwick (2002).

Another result which is valid for both West and East Germany is the substitutional relationship between high and medium skilled and the complementarity between high and low skilled employees. Regarding the relationship between low and medium skilled employees we get a differentiated picture: In West Germany the relationship is complementary while in East Germany low and medium skilled employees constitute substitutes.

Except for low skilled employees in West Germany, the own price elasticities are as expected negative. For West Germany all elasticities except for the one between high and low skilled employees is significant whereas in East Germany the elasticities are only significant on a lower level. The results reveal e.g. that in West Germany an increase of wages for medium skilled employees by 10% leads to a decrease of the labor demand by 2.8% whereas in the case of low skilled employees in East Germany this increase would lead to a decrease of the labor demand by 3.9%.

Comparing our study to earlier studies based on the same data set is difficult due to different time periods and industrial sectors that have been investigated. Bellmann, Bender, and Schank (1999) consider establishments from West Germany in 1995 and use a translog function. Further, they differentiate between blue and white collar employees and find the following own price elasticities for the first (second) group: 0.1 (-0.3) for high skilled, -0.7 (-0.9) for medium skilled and -0.7 (0.8) for low skilled employees.

Another similar study is the one conducted by Kölling and Schank (2002) who estimate a generalized Leontief cost function for West German Establishments in the period from 1994 to 1997 and differentiate between the manufacturing and services sector. They find the following own price elasticities for the first (second) sector: -0.36 (1.06) for high skilled, -0.5 (-1.06) for medium skilled and -0.57 (-2.68) for low skilled employees. Differences in the estimation results of elasticities can be explained therefore by the use of different sub-samples of establishments and employees, time periods, cost functions and econometric estimation methods.

Contrary to the results of the previous chapter, we find only weak evidence for the hypothesis of a skill biased technological change as measured by product innovations. In West Germany product innovations increase labor demand for high as well as low skilled employees, whereas they have no significant effect in East Germany. Organizational changes on the other hand reduce the demand for low skilled employees in both West and East Germany but have no impact on high skilled employees. This result is in accordance with earlier studies. Bauer and Bender (2002) show that organizational changes lead to a skill bias in favor of high skilled employees. Falk (2001) finds similar results using data from the Mannheim Innovation Panel.

Another noticeable finding is the endogeneity of the decision to introduce innovations and organizational changes which justifies the use of a simultaneous framework. We estimated the covariances between each factor demand and the two Probit equations. The problem that arises in computing covariances between regression and Probit equations is due to the fact that one of the two variables is not observable. Therefore, we have used a two-step procedure suggested by Olson (1979). In a first step the parameters of the structural form are estimated. These estimated parameters are then plugged into a bivariate likelihood function which models the joint probability of the regression and Probit equation and is assumed to depend only on the covariance between these two equations. Maximizing the function with respect to this covariance yields an estimator for the covariance. The so estimated covariances show that for West as well as for East Germany in each estimated demand equation at least one of the two included variables for innovation and organizational changes is endogenous (see tables F.7 and ?? in the appendix.) Taking account of the endogeneity of innovations and organizational changes is therefore imperative.

7.5 Summary and Conclusions

The aim of this study was the quantification of the employment effects of innovations and organizational changes and the question whether these two determinants are able to explain the increasing demand for high skilled employees. We were able to use a linked employer employee data set which combines both information on establishments with detailed individual data on the structure of the workforce.

Three formal qualification groups were differentiated and a generalized Leontief cost function estimated to derive factor demand functions. A problem thereby arises from the fact that establishments decide endogenously about the introduction of innovations and organizational changes. For this reason a mixed simultaneous equation model of type I was estimated.

The results indicated that technological change as measured by product innovations has not a clear cut impact on the labor demand for different skill groups. In West Germany they increase the labor demand not only for the group of high but also for low skilled employees while having no significant impact in East Germany. Organizational changes on the other hand reduce the demand for low skilled employees whereas high skilled employees are not affected by them. It thus seems that organizational changes are the driving force for an increasing labor demand for high skilled employees.

Chapter 8

Technological Change in an Ageing Society: Age Biased Technological Change?

8.1 Age Discrimination in the Labor Market?

*Ihr seid eine notwendige Ergänzung in einer Welt, die sich für den Schwung der Jugend und für die Kraft der sogenannten „besten" Jahre begeistert, in einer Welt, in der so sehr zählt, was man zählen kann.**

Ioannes Paulus PP. II

The previous chapters presented convincing empirical evidence that technological change is skill biased in favor of high skills. Driving forces for this increasing gap in the employment prospects between high and low skilled employees are mainly disembodied process innovations, i.e. organizational changes within establishments like e.g. decentralization and flexibilization.

An omnipresent skill biased technological change will affect all employees within a given skill group in the same way. But there are a couple of reasons why certain subgroups are expected to suffer more than others.[1] Especially older employees might be affected by technological change, because:

– Older workers are less skilled than younger ones on average.[2]

*Speech to the elderly, Munich, 19 November 1980. Own translation: You are a necessary complement in a world which is crazy about the drive of youth and the power of 'prime', a world in which only the things count which can be counted.

[1]See e.g. Friedberg (1999).

[2]In Germany e.g. in 1965 the share of individuals who finished secondary school amounted to 25%. In the mid-90s this share climbed to 37%. Cf. Engeln et al. (2003).

- High skilled older workers received their education a long time ago and due to knowledge depreciation their skills may have become obsolete in the meantime.
- Older workers have less incentives to invest in new knowledge since their time horizon until retirement is shorter as compared to younger employees.

As a consequence, an increasing technological change which requires new skill structures might harm older employees more than younger ones. Establishments will reduce their demand for older employees or older employees may decide to withdraw from the labor market and retire.

Although there is a large literature on the skill biased nature of technological change, studies on the impact on the age structure of labor demand are rather scarce, especially for Germany where at least to our knowledge not a single one exists. A recent exception for the US is Friedberg (1999) who analyzed the impact of computer use on workers of different ages. She found, after controlling for possible endogeneity effects, that the probability for employees older than 50 years to use a computer at their job decreases dramatically and at the same time they are 28% more likely to retire than their computer using counterparts.

Other studies have been conducted by Juhn (1992) and Farber (1996). Using data from the Current Population Survey, Juhn (1992) shows that the decline in employment between 1967 and 1987 in the U.S. was particularly severe among less educated male employees and suggests that the rising education premium contributed to earlier retirement among older low skilled relative to high skilled employees. Farber (1996) examines changes in the incidence of job losses in the U.S. between 1981 and 1993 using again data from the Current Population Survey. He finds that older workers have experienced more job losses than they used to, though still less than younger workers. Bartel and Sicherman (1993), finally, focused on the impact of technological change on the retirement decision of older workers between 1966 and 1983 in the US using the National Longitudinal Surveys of Older Men. They found that costs of training associated with unanticipated technological change led workers to retire earlier.

In the following we will present empirical evidence for Germany and analyze the impact of technological change on the structure of labor demand in more depth. Using an unique linked employer-employee dataset for 1993 to 1997 enables us to interact different age and skill groups therefore gaining a deeper insight into the nature of technological change. In the previous sections we have estimated a highly parameterized cost function and thereby explicitly modelled the endogeneity of technological change. In this chapter we will relax these assumptions and only rely on appropriate independence conditions using a multiple treatment framework. We will discuss the econometric issues in section three after having presented the dataset in the next section. Section four contains the estimation results while section five derives some stylized facts and concludes.

8.2 Dataset and First Descriptive Results

In the following we will briefly describe the variables used to test the skill and age biased technological change hypothesis. Again we are working with the Linked IAB Establishment Panel dataset (LIAB) which was already introduced in detail in the previous chapter. The relevant treatment variables reflecting the innovation behavior of establishments are the following:

- Product: Product innovations introduced before 1993 (yes/no), taken from the 1993 wave of the IAB Establishment Panel. Product innovations comprise the introduction of a product/service which was new for the own establishment (local innovation) or even new for the market (global innovation).
- Process: Process innovations introduced in 1992 (yes/no), taken from the 1993 wave of the IAB Establishment Panel. The dummy variable Process equals one if investments were undertaken which aimed at a reduction of labor and operations costs.
- Orga: Organizational changes introduced before 1993 (yes/no), taken from the 1995 wave of the IAB Establishment Panel. Orga is equal to one if an establishment has introduced one of the following measures: Flattening of hierarchical levels, decentralization of decision authority, introduction of group workplaces with own responsibilities, introduction of units with own costing and result calculation and introduction of just-in-time production.

The outcome variables were taken from the Employment Statistics Register at the Federal Employment Office for the years 1993 to 1997. We considered the following employment groups and combinations between them:

- Total: Total number of employees.
- High skilled: Share of employees with an university or equivalent degree.
- Medium skilled: Share of employees with an apprenticeship.

- Low skilled: Share of employees with no formal educational attainment.
- Young employees: Share of employees younger than 30 years.
- Middle employees: Share of employees [30;50) years old.
- Older employees: Share of employees older than 50 years.

Since we are especially interested in possible discrimination effects against low skilled and older employees, we will leave out medium skilled employees and those which are between 30 and 50 years old in the following analyses. Note that we are working with employment shares which sum up to one so that these shares can be deduced from the remaining information.

Combining all establishments which participated in the IAB Establishment Panel in 1993 and 1995 with employee information from 1993 to 1997 yields a total number of 2,785 West German establishments.[3] Excluding establishments from the agricultural and public sector reduces the sample size to 2,429 establishments.[4] The following figure 8.1 summarizes the assumed dependency structure where Y^T illustrates an exemplary development of the outcome variable (e.g. labor demand for high skilled employees) for treated and Y^C for control establishments.

Figure 8.1: Dependency Structure

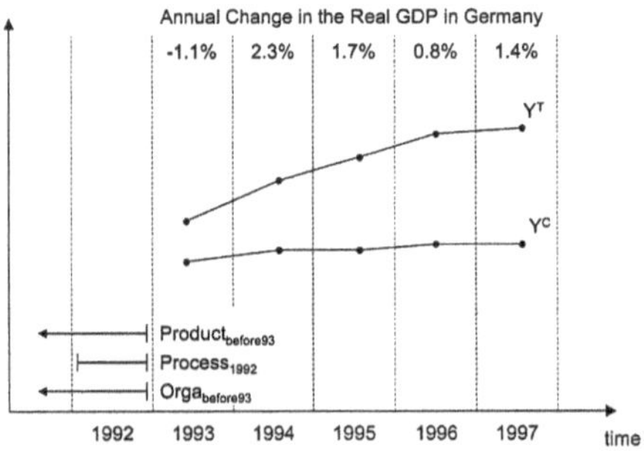

[3]Note that East-German establishments were not included until 1996 in the IAB Establishment Panel.

[4]Several comments are in order regarding this dataset. Employee information are available for the years 1993 to 1998. Due to the aforementioned structural break in the data collection methods in 1998 employment level information from this year are not directly comparable with previous years. We decided therefore to exclude them from the further analysis. Questions regarding innovation behavior were only included in the IAB survey in 1993 and then again in 1998 so that we are using those from 1993.

Before discussing appropriate econometric methods which can be used to assess the causal effects of innovation activities on labor demand we will present some simple descriptive results. Table 8.1 contains the share of establishments which have either introduced a product innovation, process innovation, organizational changes or combinations of these. One can see that most of the establishments (31%) have neither introduced a product nor a process innovation nor an organizational change. 34% of all establishments only introduced one single measure. A considerable number of establishments (25%), however, introduced two and 10% even all three measures simultaneously.

Table 8.1: Distribution of the Treatment Variable

Treatment		Number	Share
0	Nothing	749	0.31
1	Product	389	0.16
2	Process	245	0.10
3	Orga	189	0.08
4	Product & Process	269	0.11
5	Product & Orga	198	0.08
6	Process & Orga	138	0.06
7	Product, Process & Orga	252	0.10

Table 8.2 contains the means and standard deviations of the outcome variables for 1993 to 1997. One can see that the average total employment has declined during this time from 704 to 565 employees in our sample. On the other hand, the skill and age composition has remained fairly stable for high skilled and older employees and decreased by one percentage point for low skilled employees and by even six percentage points for younger employees.

In the following we will show how these outcome variables differ for establishments which have introduced product innovations, process innovations and organizational changes as compared to establishments which were not innovative at all. Figure 8.2 shows first of all that innovative establishments are on average larger than their non-innovative counterparts. In the following years both groups have experienced declining employment. However, innovative establishments were affected more strongly by this decline as can be seen by the dashed line: Between 1993 and 1997 innovative establishments on average reduced employment

Table 8.2: Descriptive Variables

	1993		1994		1995		1996		1997	
Variable	Mean	SE	Mean	SE	Mean	SE	Mean	SE	Mean	SE
Total	704	1,762	625	1,477	618	1,462	580	1,400	565	1,382
High	0.06	0.11	0.06	0.11	0.07	0.11	0.06	0.11	0.06	0.11
Low	0.27	0.23	0.27	0.23	0.26	0.22	0.26	0.22	0.26	0.23
Young	0.29	0.19	0.27	0.18	0.26	0.18	0.24	0.18	0.23	0.17
Old	0.22	0.16	0.22	0.15	0.22	0.15	0.22	0.15	0.23	0.16

by about 250 employees as compared to non-innovative ones.[5]

Figure 8.2: Simple Mean Comparison for Total Employment

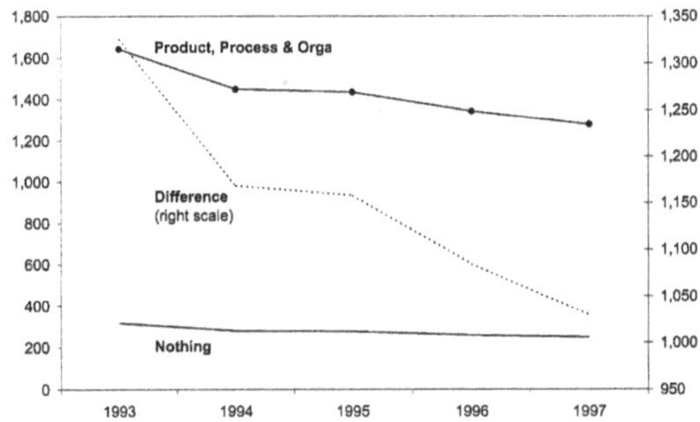

The graphs in figure 8.3 give us a first impression about the impact of innovations (measured again by the simultaneous introduction of product innovations, process innovations and organizational changes) on the labor structure. The upper left (right) figure reveals that innovations increase (reduce) the demand for high (low) skilled employees. Simple mean comparisons therefore indicate that technological change might be skill biased. Regarding different age groups the patterns are not so clear cut. We find that both groups of young and older employees are set off by technological change.[6] It is the task of the next chapters to verify whether these simple mean comparisons also reflect the true causal relationships or if these findings are simply caused by sample selection effects.

[5]The appendix to this chapter contains the same figures for other treatments.

[6]Again the appendix contains similar figures for different treatments. They reveal that different treatments have a heterogenous impact on labor demand.

Figure 8.3: Simple Mean Comparison for Different Skill and Age Groups

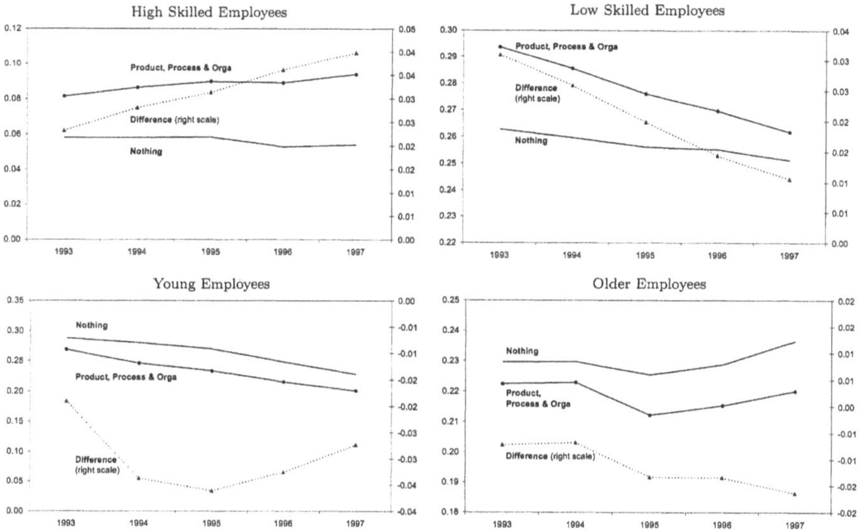

8.3 A Multiple Treatment Framework

Chapter two has introduced the basic binomial treatment framework which can be used to estimate the causal effect of a certain treatment over no treatment on a certain outcome variable. In our application, however, establishments decide whether to introduce a product innovation, a process innovation, an organizational change or to do nothing. This yields a total number of $2^3 = 8$ disjunctive treatment decisions. Table 8.1 which was presented previously contains the distribution of this treatment variable and makes clear that a large fraction of establishments (33%) has introduced more than one measure simultaneously.

Since it is reasonable to assume that the introduction of a product innovation has other employment effects than a process innovation or even the introduction of both, it makes no sense to aggregate all possible combinations into one single treatment.[7] The simple binomial setting therefore needs to be extended to a multiple treatment framework which can take this possible impact heterogeneity into account. Figure 8.4 illustrates the difference between a binomial and a multi-

[7] Cf. Plewis (2002).

ple treatment setting. In the upper part of the figure, an individual experiences a treatment, i.e. $D_i = 1$. In the post-treatment period we observe for this individual $Y_i(D_i = 1)$ and the counterfactual outcome is given by $Y_i(D_i = 0)$. If more than one treatment is possible, however, depending on the treatment we observe one of the following outcomes: $Y_i(D_i = 1), \ldots, Y_i(D_i = 7)$ and the other ones are the counterfactual outcomes.[8]

Figure 8.4: Binomial versus Multiple Treatments

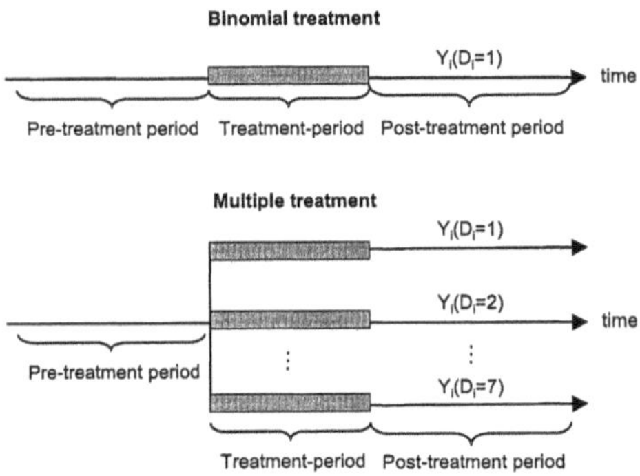

Recall once again the causal framework for a binomial treatment D. Given a pre-specified outcome variable Y the average treatment effect on the treated is given by:

$$\Delta^{ATE|T} = E_i[Y_i(1) - Y_i(0)|D_i = 1] = E_i[Y_i(1)|D_i = 1] - E_i[Y_i(0)|D_i = 1]$$

and the fundamental evaluation problem occurs because the second term $E_i[Y_i(0)|D_i = 1]$ is not observable.

Using the balancing property of the propensity score, $p(X_i)$, and imposing the conditional independence assumption $Y_i(0) \perp D_i|p(X_i)$ implies $E_i[Y_i(0)|D_i = 1, p(X_i)] = E_i[Y_i(0)|D_i = 0, p(X_i)]$ and hence the counterfactual outcome could be approximated by:

$$E_{p(X_i)|D_i=1}\left[E_i[Y_i(0)|D_i = 0, p(X_i)]\right].$$

Consider now the case when there are $m = 0, 1, \ldots, M$ mutually exclusive treatments where 0 indicates the 'no treatment' category. There are $(M+1)$ different hypothetical outcomes for each individual, namely: $\{Y_i(0), Y_i(1), \ldots, Y_i(M)\}$.

[8]More details can be found in Lechner (2001, 2002, 2002b) and Gerfin and Lechner (2002).

The fundamental evaluation problem occurs since we only observe one of these components and the remaining M outcomes are the counterfactuals.

Let $D_i \in \{0, 1, \ldots, M\}$ indicate participation in a particular treatment. Similar to the binomial case, different pair-wise treatment effects can be defined (cf. Lechner (2001, 2002)):

$$\begin{aligned}
\Delta^{ml} &= E_i[Y_i(m) - Y_i(l)] \\
\Delta^{ml|m,l} &= E_i[Y_i(m) - Y_i(l)|D_i = m, l] \\
\Delta^{ml|m} &= E_i[Y_i(m) - Y_i(l)|D_i = m].
\end{aligned}$$

Δ^{ml} gives the average treatment effect of treatment m compared to treatment l for a randomly selected individual, whereas $\Delta^{ml|m,l}$ and $\Delta^{ml|m}$ condition this effect on participants randomly selected from the group of participants in treatment m or l and only m, respectively.

In the binomial case with one treatment and one non-treatment category, only one possible pairwise comparison arises, namely treatment as compared to non-treatment. Given $M + 1$ different treatments including the non-treatment case on the other hand implies that $\binom{M+1}{2}$ different pairwise comparisons can be estimated. For $m = 1$ and $l = 0$ e.g., $\Delta^{ml|m}$ measures the employment impact for an establishment which has introduced a product innovation as compared to the non-treatment case, i.e. it answers the question whether product innovations create or destroy jobs.

Identification of $\Delta^{ml|m} = E_i[Y_i(m) - Y_i(l)|D_i = m]$ requires only identification of $E_i[Y_i(l)|D_i = m]$, whereas for the identification of $\Delta^{ml|m,l}$ one needs also the expression $E_i[Y_i(m)|D_i = l]$. To see why look at the following decomposition:

$$\begin{aligned}
\Delta^{ml|m,l} &= E_i[Y_i(m) - Y_i(l)|D_i = m, l] \\
&= p(D_i = m|D_i = m, l)E_i[Y_i(m) - Y_i(l)|D_i = m] \\
&\quad + (1 - p(D_i = m|D_i = m, l))\, E_i[Y_i(m) - Y_i(l)|D_i = l] \\
&= p(D_i = m|D_i = m, l)\Delta^{ml|m} + (1 - p(D_i = m|D_i = m, l))\, \Delta^{ml|l}
\end{aligned}$$

with $p(D_i = m|D_i = m, l) = \frac{p(D_i=m)}{p(D_i=l)+p(D_i=m)}$.

Identification of Δ^{ml} requires even the identification of all counterfactuals of $Y_i(m)$ and $Y_i(l)$ since Δ^{ml} can be decomposed according to:

$$\Delta^{ml} = \sum_{j=0}^{M} (E_i[Y(m)|D_i = j] - E_i[Y(l)|D_i = j])\, p(D_i = j) = \sum_{j=0}^{M} \Delta^{ml|j} p(D_i = j).$$

Hence, if $\Delta^{ml|j}$ is identified for all $j = 0, 1, \ldots, M$ all other effects are defined as well.

The conditional independence assumption of the binomial framework $Y_i(0), Y_i(1) \perp D_i | X_i$ can be extended to the multiple framework.[9] It is given by:

$$Y_i(0), Y_i(1), \ldots, Y_i(M) \perp D_i | X_i$$

and can also be generalized to the following independence assumption conditional on the propensity score:

$$Y_i(0), Y_i(1), \ldots, Y_i(M) \perp D_i | \left(p^1(X_i), p^2(X_i), \ldots, p^M(X_i)\right).$$

In the binomial case the propensity score is the probability of an individual to participate in a treatment. Since in case of the multiple framework we have $M+1$ different treatments the propensity score now becomes a vector of M independent probabilities $p^j(X_i) = p(D_i = j | X_i)$.[10]

There are a set of conditional independence assumptions which identify some of the possible pairwise treatment effects. The first one is the strongest and states that:

$$Y_i(m), Y_i(l) \perp D_i | \left(p^1(X_i), p^2(X_i), \ldots, p^M(X_i)\right).$$

This form of conditional independence implies that all counterfactuals are identified since for all $j = 0, 1, \ldots, M$:

$$\begin{aligned} E_i[Y_i(m)|D_i = m, p(X_i)] &= E_i[Y_i(m)|D_i = j, p(X_i)] \\ E_i[Y_i(l)|D_i = l, p(X_i)] &= E_i[Y_i(l)|D_i = j, p(X_i)] \end{aligned}$$

with $p(X_i) = \left(p^1(X_i), p^2(X_i), \ldots, p^M(X_i)\right)$. Therefore, all average treatment effects $\Delta^{ml|m}$, $\Delta^{lm|l}$, $\Delta^{ml|m,l}$ and Δ^{ml} can be identified.

The second less restrictive form of conditional independence assumption is relaxed to hold only for the subpopulation of individuals participating either in treatment m or l:

$$Y_i(m), Y_i(l) \perp D_i | \left(p(D_i = l | D_i = m, l)\right), D_i \in \{m, l\}.$$

Under this assumption only the counterfactuals $E_i[Y_i(m)|D_i = l, p(X_i)]$ and $E_i[Y_i(l)|D_i = m, p(X_i)]$ are identified using $E_i[Y_i(m)|D_i = m, p(X_i)]$ and $E_i[Y_i(l)| D_i = l, p(X_i)]$, respectively. Hence only $\Delta^{ml|m}$, $\Delta^{lm|l}$ and $\Delta^{ml|m,l}$ can be estimated.

[9] Cf. Imbens (2000).

[10] Thus if the number of elements in X is less than M it makes no sense to condition on the propensity vector since the dimensionality even increases.

The least restrictive conditional independence assumption, however, is obtained if we further relax the second one to hold only for the potential outcome $Y_i(l)$, i.e.

$$Y_i(l) \perp D_i | \left(p(D_i = l | D_i = m, l)\right), D_i \in \{m, l\}.$$

In this case only $E_i[Y_i(l)|D_i = m, p(X_i)]$ can be identified using the observable counterpart $E_i[Y_i(l)|D_i = l, p(X_i)]$. As a conclusion only $\Delta^{ml|m}$ is estimable.

In the following we will concentrate on the treatment effects $\Delta^{ml|m}$ for $l = 0$ and $m = 1, 2, \ldots, 7$ and consider thereby various outcome variables like e.g. employment shares for high and low skilled and young and elder employees, respectively (see also table 8.3). A comparison of the different treatment effects can be used to reveal complementarities. Introducing process innovations in combination with appropriate organizational changes e.g. might have larger employment effects than introducing both measures in isolation. In this case the sum of $\Delta^{2,0|2}$ and $\Delta^{3,0|3}$ will not in general be equal to $\Delta^{6,0|6}$. It is in fact the existence of such complementarities which makes the application of a multiple treatment framework necessary.[11]

Table 8.3: Considered Treatment Effects

Treatment Effect	Interpretation
$\Delta^{1,0\|1} = E_i[Y_i(1) - Y_i(0)\|D_i = 1]$	Isolated impact of only product innovations
$\Delta^{2,0\|2} = E_i[Y_i(2) - Y_i(0)\|D_i = 2]$	Isolated impact of only process innovations
$\Delta^{3,0\|3} = E_i[Y_i(3) - Y_i(0)\|D_i = 3]$	Isolated impact of only organizational changes
$\Delta^{4,0\|4} = E_i[Y_i(4) - Y_i(0)\|D_i = 4]$	Impact of Product & Process Innovations
$\Delta^{5,0\|5} = E_i[Y_i(5) - Y_i(0)\|D_i = 5]$	Impact of Product Innovations & Organizational Changes
$\Delta^{6,0\|6} = E_i[Y_i(6) - Y_i(0)\|D_i = 6]$	Impact of Process Innovations & Organizational Changes
$\Delta^{7,0\|7} = E_i[Y_i(7) - Y_i(0)\|D_i = 7]$	Impact of Product-, Process & Organizational Innovations

Recall once again how we proceeded to calculate the average treatment effect on the treated in the binomial framework once the conditional identifying assumption had been imposed. The matching estimator in this case is given by:

$$\widehat{\Delta}^{PMAT} = E_{p(X_i)|D_i=1}\left[E_i[Y_i(1)|D_i = 1, p(X_i)] - E_i[Y_i(0)|D_i = 0, p(X_i)]\right] = \Delta^{ATE|T}$$

and amounts to finding for every participating individual one non-participating individual with almost the same propensity score $p(X_i)$.

[11] Otherwise one could simply conduct three separate binomial treatment analyses for product innovations, process innovations and organizational changes.

In the multiple framework this basic idea is extended. To calculate e.g. $\Delta^{ml|m} = E_i[Y_i(m) - Y_i(l)|D_i = m]$, i.e. the average treatment effect of treatment m compared to treatment l on a randomly picked participant in treatment m, it is sufficient to impose the weakest form of conditional independence assumption, i.e. $Y_i(l) \perp D_i | \left(p(D_i = l|D_i = m, l)\right), D_i \in \{m, l\}$. Note that we have to condition only on an one-dimensional scalar $p(D_i = l|D_i = m, l)$ instead of a M-dimensional vector of all probabilities $p^1(X_i), \ldots, p^M(X_i)$. Appendix G contains some practical issues how to construct an adequate control group.

8.4 Estimation Results

The first step of every matching protocol consists of estimating the vector of M different independent probabilities $p^j(X_i) = p(D_i = j|X_i)$ for $j = 1, 2, \ldots, 7$. In our case two different approaches can be followed. The most simple one is to estimate three separate Probit models. As a result one obtains the following three probabilities: p_1 for the probability to introduce a product innovation, p_2 for process innovations and p_3 for organizational changes. These can be used to calculate the probabilities for the different treatment decisions according to:

$$\begin{aligned} p(D = 0) &= (1 - p_1)(1 - p_2)(1 - p_3) \\ p(D = 1) &= p_1(1 - p_2)(1 - p_3) \\ &\vdots \\ p(D = 7) &= p_1 p_2 p_3. \end{aligned}$$

For the estimation we have included a variety of control variables which are expected to have an impact on innovation activities and which were also discussed in chapter one. The Schumpeter size hypothesis was considered by the number and squared number of employees in 1992. Cash flow was approximated by the variable PROFIT_{93} and CAPITAL which measure profitability in 1993 (1: Very good, ..., 5: Insufficient) and the fact whether the considered establishment is a capital company or not.[12] Internal technological capabilities are measured by the following variables: STATE (state of the technological equipment, 1: Up-to-date, ..., 5: Out-of-date), ICT (investment in ICT in 1992), R&D (existence of a R&D department) and MARKET (existence of a market research department). The demand situation is reflected by the variable DEVP which measures the business

[12]Unfortunately we had no information about business development in 1992.

development between 1992 and 1993 (1: decreasing, 2: constant, 3: increasing). Due to insufficient data, technological opportunities and appropriability conditions can only be approximated by the use of industry dummies.[13] Additionally, we also included a dummy variable indicating the existence of a workers council (WORKER).[14]

Table 8.4: Estimation Results for the Probit Models

Variable	Product Innovations Parameter	Product Innovations t-Value	Process Innovations Parameter	Process Innovations t-Value	Organizational Changes Parameter	Organizational Changes t-Value
Constant	-1.40	-5.82	-1.55	-6.31	-1.66	-6.8
$EMPL_{92}$	0.00	0.97	0.00	3.80	0.00	4.77
$EMPL_{92}^2$	0.00	-0.16	0.00	-2.57	0.00	-3.04
$PROFIT_{93}$	0.01	0.38	0.10	2.92	0.06	1.69
CAPITAL	0.13	1.93	0.20	2.76	0.38	5.04
STATE	-0.02	-0.51	-0.06	-1.39	0.08	1.83
ICT	0.31	4.57	0.59	8.00	0.36	4.85
R&D	0.36	4.07	0.27	2.97	0.00	0.04
MARKET	0.24	3.10	0.14	1.79	0.04	0.57
DEVP	0.08	1.89	-0.01	-0.14	0.02	0.41
WORKER	0.07	0.97	0.21	2.64	0.38	4.68
	χ^2-Value	p-Value	χ^2-Value	p-Value	χ^2-Value	p-Value
LR-Industry	55.31	0.00	58.12	0.00	28.31	0.00
LR-Overall	249.21	0.00	501.90	0.00	328.62	0.00
Pseudo R^2	0.09		0.19		0.13	
Observations	1,945		1,945		1,945	

Table 8.4 contains the appropriate estimation results. All estimated models are highly significant according to overall likelihood ratio tests and show pseudo R^2 values ranging from 0.09 for the product innovation equation to 0.19 for process innovations. Size of the establishment is significant for the process and organizational change decision. For these equations we also find non-linear albeit small effects. As expected, a better cash flow situation has a positive impact on innovation activities. Investment in ICT has a significant and positive impact in all three equations whereas state of technology is found to have only an impact on organizational changes. The existence of a R&D and market research department increases product and process innovations. Better demand conditions as indicated

[13] We considered the following industries: Electricity and mining, primary industry, investment goods industry, consumer goods industry, construction, commerce, transportation and communication, finance and insurance, consumer services.

[14] Regional dummies as well as interaction effects turned out to be insignificant in the estimation.

by an increasing business development spur product innovations. Finally, industrial affiliation turns out to be significant in all estimated models.

Table 8.5: Estimated Marginal Effects

Variable	Product Innovations Parameter	*t*-Value	Process Innovations Parameter	*t*-Value	Organizational Changes Parameter	*t*-Value
$EMPL_{92}$	0.00	0.97	0.00	3.79	0.00	4.74
$EMPL_{92}^2$	-0.00	-0.16	-0.00	-2.57	-0.00	-3.04
$PROFIT_{93}$	0.004	0.38	0.04	2.92	0.02	1.69
CAPITAL	0.05	1.94	0.08	2.80	0.13	5.25
STATE	-0.008	-0.51	-0.02	-1.39	0.03	1.83
ICT	0.12	4.64	0.21	8.57	0.12	5.05
R&D	0.14	4.10	0.11	2.95	0.001	0.04
MARKET	0.09	3.11	0.05	1.78	0.02	0.57
DEVP	0.03	1.89	-0.002	-0.14	0.01	0.41
WORKER	0.02	0.97	0.08	2.67	0.13	4.77

Table 8.5 contains the appropriate marginal effects. The existence of a R&D department is the dominant factor for product innovation activities and increases the probability by 14% given that all other variables take on their mean values. Equally important are investments in ICT (12%) and the existence of a market research department (9%). We therefore find support for the technological capabilities hypothesis discussed in chapter one. The dummy variable indicating a capital company is also significant and increases the probability to introduce product innovations by 5%. Finally, the demand push hypothesis is also supported by the empirical evidence: An increasing business development increases the innovation probability by 3%.

Regarding the decision to introduce process innovations the most important and significant factors are investment in ICT (marginal effect: 21%), existence of a R&D department (11%), capital company (8%), market research (5%) and profitability (4%). As it was the case for the product innovation decision, internal technological capabilities play also the dominant role for process innovations. On the other hand, cash flow considerations and the demand situation have only a minor impact. Finally, organizational changes are mainly influenced by the CAPITAL dummy and the existence of a workers council which both increase the probability by 13%.

An alternative procedure for estimating the vector of the propensity scores amounts to using a multinomial model, e.g. a multinomial Logit model.[15] Looking at table 8.6 reveals that the multinomial model performs slightly better in predicting the observed shares and that the predicted probabilities using the separate Probits and the multinomial model are highly and significantly correlated (p-values < 0.00). Applying a Hausman and the Small-Hsiao test of independence of irrelevant alternatives (IIA) yields contrary results.[16] The Hausman test does not reject the IIA assumption but is plagued by a lack of positive semidefiniteness of the covariance matrix. An alternative test in this case is the one proposed by Small and Hsiao (1985) which leads to a rejection of the IIA assumption. Due to the high correlation between both sets of probabilities we should not expect different evaluation results irrespective of whether the probabilities are derived from the multinomial or the separate binomial models.[17] Since the predictive power of the multinomial Logit model is superior, we decided to base the calculation of the propensity scores on the estimation results of this model.

Table 8.6: Observed and Predicted Shares of the Treatment Decision

Treatment	Observed	Predicted Binomial	Predicted Multinomial	Correlation	IIA Test* Hausman	IIA Test* Small-Hsiao
0	0.31	0.28	0.30	0.99	–	–
1	0.16	0.17	0.16	0.88	-0.00 (1.00)	373.436 (0.00)
2	0.10	0.11	0.11	0.91	-0.00 (1.00)	349.549 (0.00)
3	0.08	0.08	0.07	0.93	-0.00 (1.00)	357.964 (0.00)
4	0.11	0.12	0.12	0.94	-0.00 (1.00)	302.811 (0.00)
5	0.08	0.07	0.08	0.88	-0.00 (1.00)	326.884 (0.00)
6	0.06	0.07	0.06	0.91	-0.00 (1.00)	239.539 (0.00)
7	0.10	0.10	0.11	0.98	-0.00 (1.00)	356.943 (0.00)

* χ^2-values for different omitted categories. p-values in brackets.

The appendix to chapter two contains the matching protocol used for the estimation of the pair-wise treatment effects given in table 8.3. Before presenting the results, several comments are in order. Imposing the common support condition

[15]Using a multinomial Probit model would circumvent the independence of irrelevant alternatives assumption. However, computational difficulties arise since in our case a seven-fold integral has to be computed.

[16]Cf. Small and Hsiao (1985).

[17]Indeed, conducting the matching using the propensity score based on the binomial Probits did not changed the results substantially.

ensures that the effects are only estimated in regions of the attribute space where observations from two treatment groups could be observed with a similar participation probability. Since we are only interested in the pair-wise treatment effects $\Delta^{ml|m}$ with $m \in \{1, 2, \ldots, 7\}$ and $l = 0$, it is sufficient to restrict the sample to the overlap for each pair.[18] Table 8.7 contains in the third (fourth) column the share of treated (control) establishments which were dropped due to the common support condition. The figures range from 13% to 30% for the group of treated and from 24% to 57% for the group of non-treated establishments.[19]

Another point concerns the fact that the matching is conducted with replacement. In a multiple treatment framework this is necessary because each treatment group will at the same time serve as a control group and hence it is not guaranteed that the number of treated will always be larger than the number of non-treated individuals. A potential consequence of matching with replacement is an increase in the variance. Looking at columns five and six of table 8.7 reveals, however, that only a small fraction of control establishments are used more than two times as a matching partner. These figures are similar in their magnitude to other studies e.g. Lechner (2002a, 2002b). Hence, matching with replacement does not seem to pose special problems.

Table 8.7: Some Matching Diagnostics

Treatment	Matched Pairs	Common Support*		Share of Controls used ...		ABIAS (%)	
		Treated	Controls	two times	> two times	Before	After
1	302	0.22	0.40	0.16	0.05	22.09	4.35
2	212	0.13	0.24	0.15	0.08	39.19	5.05
3	133	0.30	0.24	0.11	0.04	27.27	8.57
4	210	0.20	0.26	0.11	0.10	58.91	12.45
5	142	0.22	0.38	0.09	0.08	42.25	11.76
6	110	0.20	0.57	0.11	0.10	72.39	14.88
7	88	0.25	0.54	0.08	0.07	78.12	14.51

* Share of dropped establishments due to the common support condition.

[18] Cf. Gerfin and Lechner (2002), p. 868.

[19] Especially the figures for the non-treated group seem quite large and raise doubts whether differences in the treatment effects are only due to differences in the common support. Lechner (2000) proposes several robustness checks in this context, e.g. estimating the treatment effects with and without imposing the common support condition. If both effects are nearly the same, as it is the case in our application, concentrating only on the subpopulation with common support would identify the true treatment effect. Cf. Lechner (2000), p. 11.

The last point concerns the matching quality. Columns seven and eight in table 8.7 contain the average absolute standardized percentage bias (ABIAS) before and after the matching. These figures show that the differences between treated and non-treated establishments are considerably reduced after the matching.

In the following we will present the estimated treatment effects for different outcome variables Y: Total number of employees, share of high versus low skilled employees and share of young versus older employees. Finally, we will take a closer look at the share of young and older employees differentiated by skill. The treatment variables, i.e. introduction of product innovations, process innovations, organizational changes or combinations of these, refer to a time period prior to 1993 (see also figure 8.1). Information for the outcome variables on the other hand are available for the time period from 1993 to 1997. The estimated treatment effects are given by:

$$\begin{aligned}
\Delta_{94}^{ml|m} &= E_i\left[\left(Y_i^{94}(m) - Y_i^{93}(m)\right) - \left(Y_i^{94}(0) - Y_i^{93}(0)\right)|D_i = m\right] \\
\Delta_{95}^{ml|m} &= E_i\left[\left(Y_i^{95}(m) - Y_i^{93}(m)\right) - \left(Y_i^{95}(0) - Y_i^{93}(0)\right)|D_i = m\right] \\
\Delta_{96}^{ml|m} &= E_i\left[\left(Y_i^{96}(m) - Y_i^{93}(m)\right) - \left(Y_i^{96}(0) - Y_i^{93}(0)\right)|D_i = m\right] \\
\Delta_{97}^{ml|m} &= E_i\left[\left(Y_i^{97}(m) - Y_i^{93}(m)\right) - \left(Y_i^{97}(0) - Y_i^{93}(0)\right)|D_i = m\right]
\end{aligned}$$

for $m = 1, 2, \ldots, 7$.

Note that the outcome variables are measured relative to the base year 1993 instead of using level information. Applying the matching protocol on such differences has several advantages: Comparing treated with non-treated establishments for the same years ensures that changing economic conditions which have a common impact on both groups cancel out.[20] Matching on the propensity scores takes account of sample selection due to observable establishment specific covariates and avoids making some functional form assumptions. Using more than one year allows us to trace the dynamic structure of the employment effects of technological change since it can be expected that the diffusion of innovations in the market will take more than one year. And finally, by concentrating on differences of the outcome variables and neglecting level information we control for possible unobservable establishment specific effects.[21]

[20] Figure 8.1 contains the annual percentage changes in the real GDP in Germany taken from various reports of the German Council of Economic Experts. One can see that 1993 represented the end of a recession phase and that the years afterwards are characterized by positive albeit decreasing growth rates. Clearly, such changing economic conditions will also have an impact on the outcome variables.

[21] For clarification assume that $Y_{it}(m)$ is an additive separable function of an observable time and

Matching Results for Total Employment

Table 8.8 contains the estimated average treatment effects on total employment. A simple comparison between treated and non-treated establishments presented previously indicated that product innovations increase the average number of employees. Process innovations as well as organizational changes on the other hand have a negative impact on the total number of employees. The same holds true if more than one measure is implemented: Again the total number of employees decreases.

Taking account of sample selection due to observables by employing a multiple treatment framework changes these simple descriptive findings. Introducing product innovations in combination with organizational changes has now a significant employment enhancing effect.[22] Between 1995 and 1993 the total number of employees increased by 13% in treated establishments compared to their matched counterparts and remained at this level in the following years. All other treatments, however, have no significant impact on the total employment development.

Table 8.8: Matching Results for Changes in Total Employment (in %)

	1993-1994		1993-1995		1993-1996		1993-1997	
Treatment	$\Delta^{m,0\|m}$	t-Value	$\Delta^{m,0\|m}$	t-Value	$\Delta^{m,0\|m}$	t-Value	$\Delta^{m,0\|m}$	t-Value
Product	-0.06	-0.89	-0.11	-1.44	-0.06	-0.86	-0.09	-0.97
Process	0.01	0.28	-0.04	-0.62	0.00	-0.04	-0.03	-0.45
Orga	0.01	0.24	0.00	-0.07	-0.05	-0.96	0.00	-0.04
Product/Process	0.05	0.87	0.02	0.26	0.04	0.53	-0.01	-0.09
Product/Orga	0.07	1.56	0.13	2.44	0.13	2.11	0.12	1.85
Process/Orga	0.03	0.71	0.06	1.22	0.17	1.46	0.05	0.93
Prod/Proc/Orga	0.02	0.21	0.02	0.14	0.03	0.27	0.07	0.31

Matching Results for the Skill Biased Hypothesis

The estimation results contained in table 8.9 can be used to test the skill biased technological change hypothesis. The figures show that organizational changes

individual specific component x_{it} and an unobservable individual specific component ν_i: $Y_{it}(m) = x_{it} + \nu_i$. Taking differences adjusts for this unobservable heterogeneity: $Y_{it}(m) - Y_{it'}(m) = (x_{it} - x_{it'}) + (\nu_i - \nu_i) = x_{it} - x_{it'}$. For a similar proceeding see e.g. Hujer, Caliendo, and Radić (2001).

[22]We will concentrate in the following only on significant effects thereby referring to the simple $2t$ rule of thumb which implies a significance level of approximately 5%.

have a negative impact on the share of low skilled employees. Between 1993 and 1995 treated establishments employed three percentage points less low skilled employees than their matched non-treated counterparts. One year later this share decreases by another percentage point amounting then to four percentage points. However, another year later this initial negative impact has nearly vanished: We find that between 1993 and 1997 the impact on low skilled employees is still negative but not statistically significant.

On the other side, organizational changes increased the share of high skilled employees by one percentage point between 1993 and 1996 and when introduced together with process innovations again by one percentage point.[23] All other treatment effects are not significant. We therefore find support for the skill biased technological change hypothesis, at least in the medium run. However, it seems that the driving force behind this increasing wedge are not process innovations but organizational changes within establishments.[24]

Table 8.9: Matching Results for the Share of Low and High Skilled (in % points)

	1993-1994		1993-1995		1993-1996		1993-1997	
Treatment	$\Delta^{m,0\|m}$	t-Value	$\Delta^{m,0\|m}$	t-Value	$\Delta^{m,0\|m}$	t-Value	$\Delta^{m,0\|m}$	t-Value
			Low Skilled Employees					
Product	0.00	-0.24	0.00	-0.06	0.01	0.67	0.03	1.51
Process	0.01	0.89	0.00	0.04	0.00	0.12	0.01	0.38
Orga	-0.02	-1.62	-0.03	-2.23	-0.04	-2.78	-0.01	-0.97
Product/Process	0.02	0.95	0.02	0.61	0.03	1.10	0.02	0.81
Product/Orga	0.00	0.37	0.00	-0.05	0.00	-0.28	0.00	0.04
Process/Orga	0.01	0.62	-0.01	-0.67	0.00	0.12	0.00	0.05
Prod/Proc/Orga	0.00	0.12	-0.01	-0.64	0.00	-0.12	0.00	-0.17
			High Skilled Employees					
Product	0.00	0.11	0.00	0.90	0.00	0.68	0.00	0.43
Process	0.00	1.45	0.01	1.67	0.01	1.08	0.01	0.93
Orga	0.00	0.37	0.00	0.04	0.01	2.31	0.01	1.48
Product/Process	0.00	-0.32	0.00	0.71	0.01	0.63	0.00	0.70
Product/Orga	0.00	1.07	0.01	1.26	0.01	0.77	0.01	0.91
Process/Orga	0.00	0.83	0.01	1.30	0.01	1.87	0.02	1.76
Prod/Proc/Orga	0.00	0.97	0.00	0.64	0.00	0.76	0.01	1.30

[23]Note, however, that this effect is only significant at a 6% level.

[24]Indeed, although process innovations increase the share of high skilled employees between 1993 and 1995 this increase is not significant given a significance level of 5%.

Matching Results for the Age Biased Hypothesis

Besides the assumption that technological change discriminates against low skilled employees, another often raised presumption is that older employees are more often affected by technological change than younger ones. The first part of table 8.10 contains the treatment effects on employees younger than 30 years. Looking at the t-values reveals that not a single treatment has a significant impact on the development of the share of young employees. It thus seems that the labor demand for younger employees is not affected by technological progress.

Table 8.10: Matching Results for the Share of Young and Older Employees (in % points)

	1993-1994		1993-1995		1993-1996		1993-1997	
Treatment	$\Delta^{m,0\|m}$	t-Value	$\Delta^{m,0\|m}$	t-Value	$\Delta^{m,0\|m}$	t-Value	$\Delta^{m,0\|m}$	t-Value
			Young Employees					
Product	-0.01	-0.97	-0.02	-1.33	-0.02	-1.07	-0.01	-0.32
Process	0.01	0.95	0.00	0.10	-0.01	-0.31	0.00	0.09
Orga	0.00	0.37	-0.01	-0.74	-0.02	-1.25	0.02	0.89
Product/Process	-0.01	-0.54	-0.02	-0.71	-0.01	-0.41	0.01	0.31
Product/Orga	0.01	0.40	0.00	-0.27	0.00	-0.24	0.00	0.09
Process/Orga	0.01	0.55	0.00	0.12	0.00	0.18	0.01	1.13
Prod/Proc/Orga	-0.01	-0.38	-0.01	-0.32	0.00	0.08	0.01	0.32
			Older Employees					
Product	-0.02	-1.86	-0.02	-1.39	-0.03	-1.73	-0.04	-2.29
Process	-0.03	-2.20	-0.02	-1.68	-0.02	-1.16	-0.03	-1.67
Orga	0.00	-0.12	0.01	0.39	0.01	1.04	0.03	1.74
Product/Process	0.01	0.27	0.02	0.87	0.01	0.74	0.01	0.34
Product/Orga	-0.02	-1.11	-0.01	-0.91	0.02	1.37	0.02	2.03
Process/Orga	-0.01	-0.90	-0.05	-2.43	0.00	-0.19	0.00	0.15
Prod/Proc/Orga	0.00	0.04	0.00	-0.24	0.00	-0.11	0.00	0.05

The picture changes, however, for employees older than 50 years. We find evidence that product innovations reduce the share of older employees between 1993 and 1997 by four percentage points. Similar figures can be found for process innovations: After the first year the share of older employees is reduced by three percentage points and remains at this level in the following years. Process innovations which were introduced together with organizational changes reduce the share of older employees by even five percentage points between 1993 and 1995. The only positive and significant effect is found for product innovations/organizational changes with an increase of two percentage points between 1993 and 1997.

To summarize: Technological change not only discriminates against low skilled and favors high skilled employees but is also age biased. Young employees are not affected by technological change whereas the age structure has changed considerably at the expense of employees older than 50 years. The only exception is the introduction of product innovations in combination with organizational changes which has a positive impact on the share of older employees, at least in the long run.

Detailed Matching Results for Subgroups of Employees

We previously found that technological change, especially organizational changes, increases the share of high skilled employees at the expense of low skilled ones. On the other hand, technological change has no impact on young and a mixed impact on older employees: Product innovations, Process innovations and process innovations in combination with organizational changes have a negative impact while product innovations introduced together with organizational changes increase the share of older employees, at least in the long run. In the following we will go into more depth and take a closer look at the following four subgroups: Young and older low skilled and young and older high skilled employees.

Looking at the figures in table 8.11 reveals the following results: Young employees with a high qualification are again not affected by technological change as it was the case for the aggregated group of young employees. Another fact is that the share of older high skilled employees increases due to product innovations introduced together with organizational changes (however, only very modest by 0.002 percentage points between 1993 and 1995) and also due to process innovations and organizational changes between 1993 and 1997 by one percentage point but is unaffected by other treatments. On the other hand, the figures in table 8.11 also disclose the following: Not every low skilled employee is affected in the same way by technological change as suggested by the aggregated figures. Young and low skilled employees are set off by organizational changes in the magnitude of four percentage points between 1993 and 1996 whereas older low skilled employees gain (their share increases due to product innovations introduced together with organizational changes between 1993 and 1996 by one percentage point).[25]

[25]Note that is not inconsistent with the finding that the overall share of older employees is reduced. It is rather the group of older employees with a middle qualification which are responsible for the overall decrease of older employees. This follows from the fact that the shares of low, middle and high skilled older employees sum up to one.

Table 8.11: Matching Results for Subgroups of Employees (in % points)

	1994-1993		1995-1993		1996-1993		1997-1993	
Treatment	$\Delta^{m,0\|m}$	t-Value	$\Delta^{m,0\|m}$	t-Value	$\Delta^{m,0\|m}$	t-Value	$\Delta^{m,0\|m}$	t-Value
				Young & Low Skilled Employees				
Product	0.01	0.45	0.00	0.08	0.00	0.02	0.01	0.33
Process	0.02	1.46	0.01	0.65	0.01	0.57	0.01	0.86
Orga	-0.01	-1.02	-0.02	-1.99	-0.04	-2.92	-0.01	-1.02
Product/Process	0.02	0.93	0.02	0.81	0.03	1.20	0.03	1.34
Product/Orga	0.00	-0.10	0.00	-0.28	-0.01	-0.99	0.00	-0.27
Process/Orga	0.01	1.37	0.00	-0.38	0.00	-0.10	0.00	0.29
Prod/Proc/Orga	0.00	0.05	0.00	-0.29	0.00	-0.04	0.00	-0.01
				Older & Low Skilled Employees				
Product	-0.01	-1.86	-0.01	-1.59	0.00	0.01	-0.01	-0.66
Process	-0.01	-1.59	-0.01	-0.90	0.00	-0.64	-0.01	-1.16
Orga	-0.01	-1.05	0.00	0.05	0.00	-0.32	0.00	0.27
Product/Process	0.00	1.19	0.00	0.68	0.01	1.23	0.01	0.81
Product/Orga	0.00	0.98	0.01	1.32	0.01	2.56	0.01	1.70
Process/Orga	-0.01	-1.20	0.00	-0.60	0.00	-0.28	0.00	-0.94
Prod/Proc/Orga	0.00	0.29	0.00	-0.77	0.00	-0.46	0.00	-0.59
				Young & High Skilled Employees				
Product	0.00	-0.87	0.00	-1.59	0.00	-1.32	0.00	-1.63
Process	0.00	0.16	0.00	-0.01	0.00	-1.06	0.00	-0.76
Orga	0.00	1.08	0.00	-1.04	0.00	-0.73	0.00	-0.51
Product/Process	0.00	-1.04	0.00	-0.24	0.00	0.01	0.00	0.28
Product/Orga	0.00	-0.83	0.00	0.01	0.00	-0.29	0.00	0.15
Process/Orga	0.00	-0.09	0.00	-0.07	0.00	0.34	0.00	0.80
Prod/Proc/Orga	0.80	0.32	0.00	0.04	0.00	-0.12	0.00	0.35
				Older & High Skilled Employees				
Product	0.00	0.03	0.00	0.18	0.00	-0.34	0.00	-0.12
Process	0.00	-0.31	0.00	0.30	0.00	1.19	0.00	0.62
Orga	0.00	-0.56	0.00	-0.79	0.00	0.42	0.00	-0.74
Product/Process	0.00	-0.73	0.00	0.54	0.00	-0.06	0.00	0.34
Product/Orga	0.00	1.02	0.00	2.29	0.00	1.60	0.00	0.55
Process/Orga	0.00	0.08	0.00	0.78	0.00	1.62	0.01	1.86
Prod/Proc/Orga	0.00	0.89	0.00	1.19	0.00	1.09	0.00	1.37

8.5 Summary and Conclusions

In the following we will try to derive some stylized facts from the bulk of estimates that we presented previously. Figure 8.5 contains the significant treatment effects separated for two skill and two age groups and also the appropriate aggregated effects. Since the estimated treatment effects refer to employment shares it is sufficient to concentrate on the four groups in the grey boxes and the appropriate marginal distributions.[26]

Figure 8.5: Growth Rates for Different Skill and Age Groups

Skill \ Age	Young	Middle	Older	Σ_{Skill}
Low	**- 4% points** Orga (96-93)		**- 1% point** Product (94-93) **+ 1% point** Prod/Orga (96-93)	**- 4% points** Orga (96-93)
Medium				
High	No impact		**+ 1% point** Proc/Orga (97-93) **+ 0.2% points** Prod/Orga (95-93)	**+ 1% point** Orga (96-93) **+ 1% point** Proc/Orga (96-93)
Σ_{Age}	No impact		**- 4% points** — Product (96-93) **- 3% points** — Process (94-93) **- 5% points** — Proc/Orga (95-93) **+ 2% points** — Prod/Orga (97-93)	

Organizational changes were found to be the major cause for the skill biased technological change hypothesis. They reduce the share of low skilled employees by four percentage points and at the same time increase the share of high skilled employees by two percentage points (one percentage point in combination with product innovations). Thereby they unfold their impact between 1993 and 1996, i.e. within a time period of three years.

Technological change not only discriminates against low skilled but also against older employees. Process innovations reduce their share by one percentage point within one year, process innovations in combination with organizational changes by five percentage points within two years and, finally, product innovations by four percentage points within three years. The only positive effect is found for product innovations in combination with organizational changes.

Looking at the development within separate groups, 'winners' and 'losers' of technological change can be identified. 'Winners' are high skilled employees regardless of their age. Young high skilled employees and old high skilled employees are nearly unaffected while middle aged employees benefit most of all. Skill is therefore definitely a 'shield' against possible negative impacts of technological change. On the other hand, especially low skilled younger employees are the

[26]If one knows e.g. that treatment $D = 1$ increases the share of high skilled employees by 1 percentage point and decreases the share of low skilled by 0.5 percentage points, then the impact on medium skilled employees amounts to -1 + 0.5 = -0.5 percentage points.

'losers' of technological change whereas the impact on older low skilled employees is not clear cut. The employment share of young low skilled employees is reduced by four percentage points three years after the introduction of organizational changes while the share of older low skilled employees decreases by one percentage point after the introduction of product innovations and at the same time increases by one percentage point due to product innovations in combination with organizational changes.

Katz and Murphy (1992) document in their study on changes in relative wages between 1963 and 1987 in the US a pattern similar to our results. They found that older low skilled employees suffered on average less than young low skilled employees regarding their wage premium for experience. On the other hand, older high skilled employees benefited less than younger high skilled employees. An explanation for this asymmetry of skills might be skill obsolescence. Older high skilled employees received their education a long time ago which does not match the skill requirements of an accelerating technological change anymore. Given flexible labor markets like in the U.S. this qualification mismatch is reflected in decreasing relative wages while in Germany what we observe is a decreasing demand for older employees.

The existence of an age biased technological change which was documented in this chapter has several policy relevant implications. In Germany the average retirement age decreased considerably in recent years with dramatic consequences for the social systems. In 1973 e.g. the average retirement age was 62 years while in 2001 it fell below 60 years.[27] Accordingly, the labor force participation rate for employees between 55 and 64 in 2001 amounted to 36.8% compared to the OECD average of 48.4%.[28] In a recent study Herbertson and Orszag (2003) estimated the share of potential GDP loss due to early retirement and arrived at a figures of 13.2% for Germany. Decreasing labor demand for older employees therefore not only bears serious problems for the social security systems but also means a waste of resources.

Looking at the population development in Germany it is clear that the share of older employees will increase in the following years. Börsch-Supan (2002) considers different population projections and estimates that the old-age dependency ratio (the ratio of individuals over 60 years to individuals between 20 and 59) will increase from 38.6% in 1998 to values ranging from 76.7% to 88.4% in 2030.

[27]See e.g. Arnds and Bonin (2002) who also discuss potential determinants of early retirement.
[28]Cf. OECD, Labor Force Statistics, 2002.

Under such circumstances it is inevitable to encourage later retirement and therefore imperative to learn whether older employees retire because of their preferences or because of an age biased technological change. If the latter applies, i.e. if early retirement is forced by skill obsolescence like we were able to show, changes in terms of social security which aim at an increase of the retirement age will only have a limited effect. Instead, vocational training especially designed for older employees to cope with the technological progress might be in place.

Conclusions and Outlook

The aim of this thesis was to undertake an empirical analysis of different aspects of innovation activities in Germany ranging from various determinants to the impacts on the labor market. We started with an assessment of the effectiveness of public R&D-subsidies on the innovation propensity. Despite the considerable amount of money spent in this area and tight public budgets, empirical evidence for Germany is still rather scarce. By applying various parametrical and non-parametrical methods, we were able to add some new insights to this ongoing debate. Contrary to previous studies we find effects which are – except for East German establishments – at most insignificant. Further on, our results point to the view that the relation between innovation activities and granted R&D-subsidies is plagued by sample selection and that controlling only for observable covariates is not sufficient to remove all the endogeneity. Although our findings face several drawbacks, they point to the view that a demand side oriented R&D-policy might not yield an increase of innovations. Instead, improving supply side conditions, e.g. by raising the share of scientists and engineers, might be more fruitful.[29]

Another still unexplored area which was addressed in this thesis is the relation between technological and organizational changes which have been introduced recently by establishments such as flattening of hierarchical levels and decentralisation of decision authority. Although a number of studies have analyzed the impact of such organizational changes on productivity, there are hardly any studies which focus on the impact on innovations and – at least to our knowledge and except for some anecdotic evidence – not a single one for Germany. In this thesis we were able to show that organizational changes exhibit not only a statistically significant impact on the innovation propensity of establishments but that this impact is also of considerable economic importance. Additionally, we revealed that such organizational changes are more effective when they are flanked by appropriate human resource management practices.

[29] For a more thorough discussion see e.g. Romer (2000).

Organizational changes not only influences the productivity and innovation behavior of establishments but also their labor demand. In the last part of this thesis we presented empirical evidence that besides technological change organizational changes also discriminate against low skilled employees. They are therefore promising candidates to explain the recent labor market trends and might also contribute to the still unresolved productivity paradox of ICT. For these reasons one chapter of this thesis was devoted to the explanation of their recent appearance. Based on a model proposed by Lindbeck and Snower (2000) and using an unique linked employer employee dataset for Germany we were able to show that ICT trigger organizational changes and that it is in fact this complementary relation which unfolds productivity and innovation enhancing impacts and also shifts the skill requirements.

There is a large literature on the skill biased nature of technological change and a couple of studies which focus on organizational changes. Studies, however, which analyze the impact of technological and organizational changes on the age structure of labor demand are scarce, especially for Germany where to our knowledge there is not a single one. The last chapter of this thesis was intended to fill this gap. Applying a multiple treatment framework we were able to show that albeit skill 'shields' against technological change at the same time older high skilled employees benefit less than younger ones. This asymmetry of skills points to the existence of skill obsolescence which can only be counteracted by continuous efforts to refresh human capital.

We firmly belief that innovations are crucial and that innovations can be, at least to some extent, explained by economics. This was indeed the 'credo' for writing this thesis. However, there are still a lot of open questions but also several promising research paths which could be followed in the future. The approach of this thesis was merely an empirical one. Such an approach is justified since most of the datasets which can be used in this context, especially establishment level data, were not available until recently. But at the same time every econometric analysis needs to be complemented by economic theory. To this aim different economic areas should be brought together, e.g. insights from labor economics and industrial organization but also different fields of business administration such as personnel management.

Organizational changes were identified in this thesis as one of the major forces driving labor demand which also exhibit a close relation to innovations. We are convinced that organizational changes will have a lasting and subtle impact in the future as well. However, our knowledge about the forces driving them and the channels through which they operate is still limited and bears great potential for further research.

Another important issue concerns the fact that all the existing theoretical and empirical work in this field applies solely to advanced economies. The role of technological change in developing and transition countries, however, might differ substantially from these findings. With the enlargement of the European Union it will become inevitable to learn more about the innovation activities in candidate countries in order to foster the innovation capacity of both current and new member states. Fortunately, more and more candidate countries recently have undertaken efforts to improve their knowledge about their national innovation systems. While finishing this thesis, in Croatia e.g. a large survey is being prepared which will form a part of the Community Innovation Survey.

In the past technological change was associated with the fear of a 'jobless' growth and a 'de-qualification' of the workforce. One popular proponent of such a pessimistic view is e.g. Rifkin (1995). Nowadays, the opposite seems to be true. We are facing a technological change which requires higher skills and demands a greater deal of flexibility from all of us.[30] Since the multidimensional nature of technological and especially organizational changes makes it hard to assess which skills will be needed in the future, and taking into account that improvements in education and training will actually take a long time to be reflected in labor market outcomes, it is questionable if this education-innovation race can always be won. The most reliable prediction in this context seems therefore, that not only opportunities and chances offered by technological change will increase but also the uncertainties associated with them.

[30] For a popular opinion on this view see e.g. Reich (2000).

Appendix A

Additional Material to Chapter 2

A.1 Implementation of the Matching Estimator

Matching estimation amounts to construct a control group consisting of non-treated individuals which are similar to the group of treated individuals. One obvious approach to construct such a control group amounts to finding for every treated individual a control individual out of the group of non-treated individuals with exactly the same covariates, i.e. conduct an exact cell matching. However, the so called curse of dimensionality is likely to hinder such a strategy. Consider e.g. the case where we have K covariates which are all dichotomous. In this case matching on all covariates is hardly possible, because an increase in the variables increases the number of matching cells exponentially (2^K) while the number of observations remains constant. With increasing K it becomes more and more difficult to find an identical twin-establishment which could be paired with the participating one.

The dimensionality problem can be avoided by matching not on the covariates but on a balancing score $b(X)$ which is an injective function of X. It can be shown that:

$$Y_i(1), Y_i(0) \perp D_i | X_i \Rightarrow Y_i(1), Y_i(0) \perp D_i | b(X_i)$$

that is, if sample selection is ignorable given X then sample selection is also ignorable if we condition on $b(X)$. In fact if we want to identify only the average treatment effect on the treated, the following conditional mean independence assumption suffices: $Y_i(0) \perp D_i | b(X_i)$.

One possible balancing score is the propensity score, $p(X)$, of an individual. The propensity score is the conditional probability of receiving the treatment given the variables X:

$$p(X_i) \equiv p(D_i = 1|X_i) = E[D_i|X_i]$$

and can be estimated using a simple binomial probit or logit model.

Using the conditional independence assumption on the propensity score we can now match treated and control individuals on the basis of this one-dimensional variable $p(X_i)$ instead of the multi-dimensional vector X_i. From $Y_i(1), Y_i(0) \perp D_i|X_i$ it follows:

$$E_i[Y_i(0)|D_i = 0, p(X_i)] = E_i[Y_i(0)|D_i = 1, p(X_i)] = E_i[Y_i(0), p(X_i)]$$

and hence given a certain value of $p(X_i)$ we can estimate the conditional average treatment effect on the treated using the propensity score matching-estimator according to:

$$\begin{aligned} \widehat{\Delta}^{PMAT|p(X)} &= E_i[Y_i(1)|D_i = 1, p(X_i)] - E_i[Y_i(0)|D_i = 0, p(X_i)] \\ &= E_i[Y_i(1)|D_i = 1, p(X_i)] - E_i[Y_i(0)|D_i = 1, p(X_i)] \\ &= \Delta^{ATE|T,p(X)} \end{aligned}$$

where the second equality follows immediately from the conditional independence assumption. Assuming $0 < p(D_i = 1|X_i) < 1$ and taking the expectation over the distribution of $p(X_i)$ yields finally the unconditional propensity score matching-estimator of the average treatment effect on the treated:

$$\widehat{\Delta}^{PMAT} = E_{p(X_i)|D_i=1}\left[E_i[Y_i(1)|D_i = 1, p(X_i)] - E_i[Y_i(0)|D_i = 0, p(X_i)]\right] = \Delta^{ATE|T}.$$

To implement the estimation strategy suggested by the propensity score matching, four steps are necessary:

1. Estimating the propensity score: The quality of the matching estimator heavily depends on the quality of the estimated propensity score and whether the balancing property is fulfilled.

2. Matching treated with similar non-treated individuals: The second step concerns the question how to construct a control group, i.e. how to define similarity between establishments. Several procedures have been proposed to this aim.

3. Calculating the average treatment effect on the treated: The difference in the means of the outcome-variable for treated and matched control individuals may serve as an estimator of the average treatment effect on the treated.

4. Conducting some diagnostics: Propensity score matching is based on a careful construction of the control group. The similarity between treated and matched controls should be checked afterwards.

1. Estimating the propensity score

Since the 'true' propensity score is unknown, it has to be estimated using a binomial logit or probit model. These models do not need a behavioral justification. The only goal is to ensure the balancing property of the propensity score, i.e. on average treated and control individuals should have the same distribution of X.

One can also use the predicted linear index instead of the predicted probability (unbounded propensity score). The advantage of such a proceeding is that individuals in the tail of the distribution of $p(D_i = 1|X_i)$ can be better distinguished.

Ideally we would like the distributions of the propensity score to be the same for the group of treated and controls. In this case more similar controls could be found. A first way to improve the matching quality is to impose the so called common support restriction, i.e. drop those controls that have a propensity score lower than the minimum or higher than the maximum propensity score of the treated individuals and vice versa. Another way consists in matching not only on the propensity score but additionally on some covariates, like e.g. sex or age groups.

2. Matching treated with similar non-treated individuals

Several procedures for matching on the propensity score have been suggested. In general, we estimate the average treatment effect on the treated by contrasting the mean outcome of the treated individuals with a weighted average of control group observations according to:

$$\widehat{\Delta}^{PMAT} = \frac{1}{N_1}\sum_{i=1}^{N_1}\left[(Y_i(1)|D_i = 1, p(X_i)) - \sum_{j\in\{D_j=0\}} w(i,j)\,(Y_j(0)|D_j = 0, p(X_j))\right]$$

where $w(i,j)$ represents the weight attached to the j-th non-participating individual. The more similar the i-th and j-th establishments are in terms of their propensity scores, the higher should be the weight attached to the j-th establishment. Different matching estimators differ in the weights attached to the controls:

- Nearest neighbor matching: A traditional and easily to implement estimator is a pairwise matching where for every participating individual we choose that out of the group of non-participating which is closest regarding its propensity score.

- Caliper matching: One disadvantage of the nearest-neighbor-matching is that it faces the risk of bad matches if the closest neighbor is still far away. This can be avoided by imposing a tolerance level ϵ on the maximum distance between the propensity scores.
- Radius matching: A variant of the caliper matching is the radius matching where not only the nearest non-treated individual within the tolerance level ϵ is chosen as the control individual but the mean outcome of all non-treated individuals within a radius ϵ.
- Stratification or interval matching: In this variant the common support of the bounded or unbounded propensity score is partitioned into a set of intervals. Within each interval, a separate treatment effect is estimated by taking the mean difference in outcomes between treated and non-treated individuals within the interval. In order to obtain an estimate for the average treatment effect on the treated, a weighted average of the separate effects is calculated with the fraction of $D_i = 1$ observations within each interval as the weights.
- Kernel matching: One disadvantage of the caliper and radius matching is the difficulty to determine a reasonable tolerance level ϵ a priori. Furthermore, it has also been suggested to use more than one non-treated establishment to construct the control group. Thereby establishments which are more dissimilar are downweighted while higher weights are attached to those establishments which are more similar to the treated ones. As a weighting function one can use e.g. the Gaussian kernel.

Matching with replacement means that non-participating individuals may be used more than once as a matching partner whereas in matching without replacement every non-treated individual could serve as a match for at most one treated individual. Thereby a trade-off between an increased variance and an increased bias arises:

- Allowing replacement increases the average quality of the matches since the range of potential matching partners is wider therefore leading to a decreasing bias.
- On the other hand it reduces the number of independent observations used to construct the counterfactual mean, thus increasing the estimator's variance.

3. Calculating the average treatment effect on the treated

The difference in the means of the outcome-variable for the treated and the matched control individuals may serve as an approximation of the average treatment effect on the treated. Looking once again at this estimator one can see that the estimated effect is the difference of two means so that the variance of $\widehat{\Delta}^{PMAT}$ is simply the sum of the variance of the outcome-variable of the treated and matched controls. A statistical test on the significance of this difference can therefore be based on a simple t-test.

However, when matching with replacement is performed, this variance has to be corrected according to:

$$Var\left[\widehat{\Delta}^{PMAT}\right] = \frac{1}{N_1} Var\left[Y_i(1)|D_i = 1\right] + \frac{\sum_{j=1}^{N_0} \omega_j^2}{(N_1)^2} Var\left[Y_i(0)|D_i = 0\right]$$

where N_1 (N_0) is the number of treated (non-treated) individuals and ω_j^2 the number of times that a non-treated individual is used as a matched control. As can be seen the variance of the matching estimator increases with ω_j^2. When a kernel based matching is performed the variance has to be calculated using bootstrap methods.

4. Conducting some diagnostics

The reliability of all matching estimators heavily depends on the balancing property of the propensity score. A simple and intuitive test whether the matching approach is able to balance the observable covariates between the group of treated and non-treated individuals therefore consists of looking at the distribution of the covariates after the matching.

Rosenbaum and Rubin (1985) proposed to this aim the absolute standardized bias which is defined for a single covariate X_k as follows:

$$\left\| \frac{\overline{X}_k(1) - \overline{X}_k(0)}{\sqrt{\left(Var[X_k(1)] + Var[X_k(0)]\right)/2}} \right\|$$

where $\overline{X}_k(1)$ ($Var[X_k(1)]$) is the mean (variance) of the covariate in the group of treated and $\overline{X}_k(0)$ ($Var[X_k(0)]$) the mean (variance) in the group of non-treated establishments.

Having presented different matching protocols, a natural arising question is which one to choose? Smith and Todd (2003) as well as a number of other studies have shown that the choice between different matching estimators as well as the choice of the bandwidth parameter for the kernel matching do not make a big difference and that the estimated average treatment effects are robust regarding different matching procedures.

Appendix B

Additional Material to Chapter 3

B.1 Justification of the Average Treatment Effect on the Treated

The average treatment effect on the treated plays a prominent role in the practical evaluation literature. But is it also a parameter of interest for us, i.e. can its use be justified in our application? To this aim let us assume that the government has control over the policy variable φ which defines the scope of participants, i.e. φ is related to the number of subsidized as well as to non-subsidized establishments.[1] If N_1 is the number of subsidized, N_0 the number of non-subsidized and $\overline{N}$ the total number of establishments, then $N_1(\varphi) + N_0(\varphi) = \overline{N}(\varphi)$.

Assume now that we can abstract from general equilibrium effects, i.e. we neglect that the decision about the scope of participating firms, φ, might have an impact on the outcome of a single firm.[2] This assumption seems justifiable in our case since the amount of all public R&D-subsidies in Germany in 2000 only made up 5.2% of total R&D-spending. As a consequence, the expected effect of public subsidies at level φ on the level of total macroeconomic innovation activities, $E[Y(\varphi)]$, is given by:

$$E[Y(\varphi)] = N_1(\varphi)E[Y_i(1)|D_i = 1] + N_0(\varphi)E[Y_i(0)|D_i = 0] \quad \text{(B.1)}$$

and the optimum level of φ is determined by the first derivative of (B.1) with

[1] We abstract at this point from costs associated with the implementation and conduction of such programs. For more details see Heckman et. al. (1999).

[2] We will also abstain from any selection mechanism, i.e. abstract from any covariates that might play a role in the determination of the outcome variable. This will not alter the general conclusion of the following considerations but only facilitate the notation.

respect to φ:

$$\text{(B.2)} \qquad \frac{\partial E[Y(\varphi)]}{\partial \varphi} = \frac{\partial N_1(\varphi)}{\partial \varphi} \{E[Y_i(1)|D_i = 1] - E[Y_i(0)|D_i = 0]\}.$$

The expression inside the bracket can be estimated by taking the mean differences between the output of subsidized and non-subsidized establishments, i.e by the use of Δ^{ATE}. But what about the average treatment effect on the treated? In the following we will derive what the average treatment effect on the treated actually measures and whether this parameter is of any interest for us.

The average treatment effect on the treated reflects in our case the net effect on the innovation activities of the whole economy from the existence of a program at level $\overline{\varphi}$ compared to the alternative of shutting down all programs:

$$\text{(B.3)} \; E[Y(\overline{\varphi})] - E[Y(0)] = N_1(\overline{\varphi})E[Y_i(1)|D_i = 1] + N_0(\overline{\varphi})E[Y_i(0)|D_i = 0] \\ -N_1(0)E[Y_i(1)|D_i = 1] - N_0(0)E[Y_i(0)|D_i = 0].$$

In this sense it is an 'all or nothing' evaluation.[3] Since $N_1(0) = 0$, $N_0(0) = \overline{N}$ and $N_0(\varphi) - \overline{N} = N_1(\varphi)$, equation (B.3) can be rewritten as:

$$\text{(B.4)} \qquad E[Y(\overline{\varphi})] - E[Y(0)] = N_1(\varphi)\left(E[Y_i(1)|D_i = 1] - E[Y_i(0)|D_i = 0]\right).$$

Assuming now that the decision to participate in an R&D subsidy depends on φ, Heckman et. al. (1999) have shown that at the optimum level the marginal effect from participation is equal to non-participation which implies $E[Y_i(0)|D_i = 1] = E[Y_i(0)|D_i = 0]$. Hence the result of the 'all or nothing' evaluation can be stated as:

$$\text{(B.5)} \quad N_1(\varphi)\left(E(Y_i(1)|D_i = 1) - E(Y_i(0)|D_i = 1)\right) = N_1(\varphi)\Delta^{ATE|D}$$

which gives us finally a justification of the average treatment effect on the treated: If we can abstain from general equilibrium effect and assume that the utility of the marginal entrant into the model is zero, the average treatment effect on the treated gives us the net effect of the program on the outcome variable as compared to the alternative scenario of no program.[4] If it is positive, conducting the program is efficient. However, we have to keep in mind that $\Delta^{ATE|D}$ says nothing about the optimal level φ.

[3]See Heckman et.al. (1999).

[4]In all other cases, however, as Heckman et. al. (1999) stated, the average treatment effect on the treated is inappropriate for evaluating the effect of a policy change.

B.2 Variables Used for the Innovation Outcome Equation

There is a vast literature about the determinants of innovations.[5] A useful classification is a differentiation between firm and market specific factors. Market factors include e.g. the competition intensity, market concentration, exposure to international trade but also demand factors, like profitability and expected development of the business volume which are all expected to have an innovation enhancing impact (see also table B.1).

Table B.1: Variables Used for the Estimation of the Outcome Equation

Variable	Mean	SE
Competition intensity in 1998 (1 = No pressure, ..., 4 = High pressure)	3.43	0.83
Gini concentration of business volume in 1998	0.82	0.06
Export share in 1998	6.05	16.85
State of technology used in 1998 (1 = Up-to-date, ..., 5 = Out-of-date)	2.12	0.79
R&D-department existing	0.14	0.35
Share of high qualified employees in 1998	0.62	0.42
Number of R&D co-operations	0.34	0.99
Number of employees in 1998	252	1,174
Share of one man businesses	0.33	0.47
Share of establishments organized as a business partnership	0.10	0.29
Share of private limited companies	0.39	0.49
Share of capital companies	0.07	0.25
Business development in 1998 (1 = Very good, ..., 5 = Insufficient)	1.98	0.72

Firm specific factors on their part can be further split into internal and external technological capabilities. The idea behind this classification is the following: Industry sectors are characterized by differing technological opportunities which have also an impact on the innovation behavior of establishments. Additionally, innovations can be boosted if establishments cooperate with other institutions. However, in order to benefit from such external cooperations, there must be some technological expertise within the establishment. External technological capabilities may be captured by variables indicating R&D-cooperations with other institutions like universities. Internal technological capabilities include e.g. the state of technology, the existence of an R&D department, the share of high qualified employees and employees devoted to R&D. Other firm specific factors which were also included in the estimation are the size of establishments measured by the number and squared number of employees, industrial and regional dummies.

[5]For an overview see e.g. Cohen (1998) and also chapter one of this thesis.

B.3 Calculation of the Standard Errors Using the Delta Method

In the following we will briefly describe how the standard errors of (3.12) and (3.13) have been calculated. To this aim note that both expressions, i.e. the marginal effect $\frac{\partial Y_i}{\partial D_i}$ and the average treatment effect on the treated ATE^{Probit} are nonlinear functions of the estimated parameters $\widehat{\beta}$ and $\widehat{\Delta}$. Let us collect them in the vector θ, i.e. $\theta = \begin{pmatrix} \widehat{\beta}' & \widehat{\Delta} \end{pmatrix}$. Estimating β and Δ yields also the corresponding covariance matrix of θ, however, not automatically the standard error for a nonlinear function $G(\theta)$, like e.g. (3.12) and (3.13).

The delta method states that the variance of $G(\theta)$ can be approximated using a first order Taylor series:

$$Var[G(\theta)] = \nabla_\theta G(\theta)' Var[\theta] \nabla_\theta G(\theta) \tag{B.6}$$

with $\nabla_\theta G(\theta)$ as the gradient of G with respect to θ.

For the marginal effect given in (3.12), $\nabla_\theta G(\theta)$ is given by:

$$\nabla_\theta G(\theta) = \begin{pmatrix} \phi(\widehat{\beta}'X_i + \widehat{\Delta}) - \phi(\widehat{\beta}'X_i) \\ X_{i1}\left[\phi(\widehat{\beta}'X_i + \widehat{\Delta}) - \phi(\widehat{\beta}'X_i)\right] \\ \vdots \\ X_{iK}\left[\phi(\widehat{\beta}'X_i + \widehat{\Delta}) - \phi(\widehat{\beta}'X_i)\right] \\ \phi(\widehat{\beta}'X_i + \widehat{\Delta}) \end{pmatrix} \tag{B.7}$$

whereas for (3.13), $\nabla_\theta G(\theta)$ becomes:

$$\nabla_\theta G(\theta) = \begin{pmatrix} N^{-1}\sum_{i=1}^N \left[\phi(\widehat{\beta}'X_i + \widehat{\Delta}) - \phi(\widehat{\beta}'X_i)\right] \\ N^{-1}\sum_{i=1}^N \left[X_{i1}\left(\phi(\widehat{\beta}'X_i + \widehat{\Delta}) - \phi(\widehat{\beta}'X_i)\right)\right] \\ \vdots \\ N^{-1}\sum_{i=1}^N \left[X_{iK}\left(\phi(\widehat{\beta}'X_i + \widehat{\Delta}) - \phi(\widehat{\beta}'X_i)\right)\right] \\ N^{-1}\sum_{i=1}^N \left[\phi(\widehat{\beta}'X_i + \widehat{\Delta})\right] \end{pmatrix}. \tag{B.8}$$

B.4 Tables and Figures

Table B.2: Matching Estimator of the Average Treatment Effect of the Treated

	West German Establishments					
	Broad Innovation Concept			Narrow Innovation Concept		
	ATE	SE	t-Value	ATE	SE	t-Value
One-to-One Matching	0.00	0.08	-0.01	0.10	0.08	1.25
Conditional One-to-One	0.18	0.09	2.04	0.21	0.08	2.67
Kernel Matching	0.09	0.05	1.72	0.17	0.06	2.71
Conditional Kernel	0.15	0.07	2.29	0.17	0.07	2.26
	East German Establishments					
	Broad Innovation Concept			Narrow Innovation Concept		
	ATE	SE	t-Value	ATE	SE	t-Value
One-to-One Matching	0.02	0.05	0.35	0.06	0.04	1.29
Conditional One-to-One	0.11	0.05	2.29	0.12	0.04	2.82
Kernel Matching	0.09	0.03	3.33	0.11	0.03	3.67
Conditional Kernel	0.11	0.04	2.89	0.12	0.03	3.99

Table B.3: Matching Estimator of the Average Treatment Effect of the Treated (Cont.)

	Micro Establishments					
	Broad Innovation Concept			Narrow Innovation Concept		
	ATE	SE	t-Value	ATE	SE	t-Value
One-to-One Matching	0.05	0.07	0.83	0.04	0.06	0.74
Kernel Matching	0.09	0.05	1.97	0.10	0.05	2.13
	Small and Medium Establishments					
	Broad Innovation Concept			Narrow Innovation Concept		
	ATE	SE	t-Value	ATE	SE	t-Value
One-to-One Matching	0.16	0.06	2.70	0.17	0.05	3.40
Kernel Matching	0.07	0.04	1.94	0.14	0.04	3.19
	Large Establishments					
	Broad Innovation Concept			Narrow Innovation Concept		
	ATE	SE	t-Value	ATE	SE	t-Value
One-to-One Matching	-0.18	0.12	-1.58	0.01	0.11	0.12
Kernel Matching	-0.08	0.08	-1.10	0.07	0.08	0.88

Table B.4: Logit Estimation of the Treatment Decision

	All Est.		West Germany		East Germany		Micro Est.		SME		Large Est.	
	χ^2	p-Value	χ^2	p-Value	χ^2	p-Value	χ^2	p-Value	χ^2	p-Value	χ^2	p-Value
LR-Overall	446.79	0.00	93.51	0.00	113.81	0.00	162.25	0.00	233.81	0.00	78.94	0.00
LR-Industry	31.38	0.00	15.07	0.02	21.31	0.00	14.63	0.02	10.05	0.12	5.20	0.39
LR-Region	231.30	0.00	6.49	0.59	5.80	0.33	67.87	0.00	120.88	0.00	40.76	0.00
LR-Controls	96.62	0.00	39.94	0.00	60.74	0.00	19.43	0.02	41.32	0.00	21.28	0.02
Pseudo R^2	0.21		0.14		0.09		0.22		0.24		0.21	
Hit-Rate	0.73		0.81		0.63		0.73		0.70		0.62	
Naive Hit-Rate	0.71		0.86		0.57		0.76		0.69		0.63	
Observations	2,297		1,331		966		897		966		333	
	Mean	SE	Mean	SE	Mean	SE	Mean	SE	Mean	SE	Mean	SE
$\widehat{p}(D_i = 1 \mid x_i)$	0.18	0.17	0.07	0.08	0.33	0.16	0.15	0.15	0.20	0.20	0.25	0.21

Table B.5: Probit Estimation of the Innovation Decision

	Broad Innovation Concept											
	All Establishments		West Germany		East Germany		Micro Est.		SME		Large Est.	
	Estimate	SE	Estimate	SE	Estimate	SE	Estimate	SE	Estimate	SE	Estimate	SE
Δ	0.07	0.09	-0.10	0.18	0.20	0.11	0.11	0.15	0.13	0.14	-0.52	0.30
Marg. Effect	0.07	0.09	-0.03	0.06	0.08	0.04	0.04	0.05	0.04	0.05	-0.15	0.09
ATE	0.02	0.03	-0.03	0.05	0.06	0.03	0.02	0.04	0.04	0.04	-0.11	0.08
	χ^2	p-Value	χ^2	p-Value	χ^2	p-Value	χ^2	p-Value	χ^2	p-Value	χ^2	p-Value
LR-Overall	554.66	0.00	429.01	0.00	140.81	0.00	156.35	0.00	182.56	0.00	145.15	0.00
LR-Industry	69.55	0.00	36.35	0.00	39.98	0.00	55.99	0.00	23.28	0.11	20.61	0.04
LR-Region	37.27	0.00	16.06	0.10	21.92	0.00	38.01	0.00	20.68	0.19	6.73	0.92
LR-Controls	207.81	0.00	161.26	0.00	53.08	0.00	58.20	0.00	78.30	0.00	29.24	0.02
Observations	2,013		1,172		837		860		852		281	

	Narrow Innovation Concept											
	All Establishments		West Germany		East Germany		Micro Est.		SME		Large Est.	
	Estimate	SE	Estimate	SE	Estimate	SE	Estimate	SE	Estimate	SE	Estimate	SE
Δ	0.28	0.09	0.28	0.16	0.34	0.12	0.17	0.17	0.41	0.15	0.22	0.22
Marg. Effect	0.09	0.03	0.09	0.05	0.11	0.04	0.04	0.04	0.13	0.06	0.08	0.08
ATE	0.06	0.02	0.07	0.04	0.07	0.03	0.03	0.03	0.09	0.04	0.07	0.06
	χ^2	p-Value	χ^2	p-Value	χ^2	p-Value	χ^2	p-Value	χ^2	p-Value	χ^2	p-Value
LR-Overall	219.29	0.00	128.68	0.00	124.04	0.00	107.66	0.00	103.37	0.00	46.79	0.32
LR-Industry	44.41	0.00	19.15	0.26	177.77	0.00	51.40	0.00	40.70	0.00	11.60	0.39
LR-Region	19.11	0.26	7.92	0.64	6.39	0.17	19.74	0.14	11.53	0.78	7.79	0.93
LR-Controls	83.71	0.00	54.42	0.00	41.00	0.00	44.68	0.00	36.42	0.00	19.89	0.18
Observations	2,012		1,172		837		844		846		283	

Table B.6: Simultaneous Probit Estimation of the Innovation Decision

	Broad Innovation Concept									
	All Establishments		West Germany		East Germany		Micro Establishments		SME	
	Estimate	SE	Estimate	SE	Estimate	SE	Estimate	SE	Estimate	SE
Δ	-0.76	0.31	-1.59	0.33	1.23	0.46	-0.68	0.83	-1.18	0.27
Marg. Effect	-0.20	0.06	-0.32	0.05	0.46	0.15	-0.16	0.16	-0.22	0.06
ATE	-0.20	0.07	-0.32	0.04	0.39	0.15	-0.13	0.16	-0.29	0.10
ρ	0.51	0.18	0.81	0.17	-0.61	0.29	0.49	0.50	0.80	0.15
	χ^2-Value	p-Value	χ^2-Value	p-Value	χ^2-Value	p-Value	χ^2-Value	p-Value	χ^2-Value	p-Value
LR-Overall	372.36	0.00	245.60	0.00	95.02	0.00	122.33	0.00	156.15	0.00
LR-Industry	61.16	0.00	33.78	0.01	42.60	0.00	24.57	0.10	21.97	0.11
LR-Region	41.46	0.00	15.56	0.11	14.67	0.01	42.54	0.00	32.28	0.01
LR-Controls	209.60	0.00	151.07	0.00	43.31	0.00	50.55	0.00	91.84	0.00
Observations	1,970		1,150		820		851		843	
	Narrow Innovation Concept									
	All Establishments		West Germany		East Germany		Micro Establishments		SME	
	Estimate	SE	Estimate	SE	Estimate	SE	Estimate	SE	Estimate	SE
Δ	-0.78	0.49	-0.84	0.90	1.11	0.83	-1.03	0.47	-0.89	0.07
Marg. Effect	-0.15	0.06	-0.16	0.11	0.39	0.31	-0.11	0.06	-0.15	0.04
ATE	-0.17	0.09	-0.17	0.12	0.31	0.26	-0.15	1.00	-0.19	0.01
ρ	0.64	0.28	0.62	0.50	-0.45	0.51	0.74	0.28	0.83	0.06
	χ^2-Value	p-Value	χ^2-Value	p-Value	χ^2-Value	p-Value	χ^2-Value	p-Value	χ^2-Value	p-Value
LR-Overall	216.44	0.00	68.91	0.00	80.73	0.00	97.83	0.00	105.88	0.00
LR-Industry	41.25	0.00	127.56	0.00	30.50	0.01	29.24	0.02	20.14	0.21
LR-Region	24.26	0.08	23.26	0.11	3.86	0.43	27.13	0.04	30.96	0.01
LR-Controls	104.16	0.00	7.05	0.72	39.99	0.00	36.72	0.00	49.18	0.00
Observations	1,969		1,150		819		850		843	

Figure B.1: Unbounded Propensity Score for Treated and Control Establishments

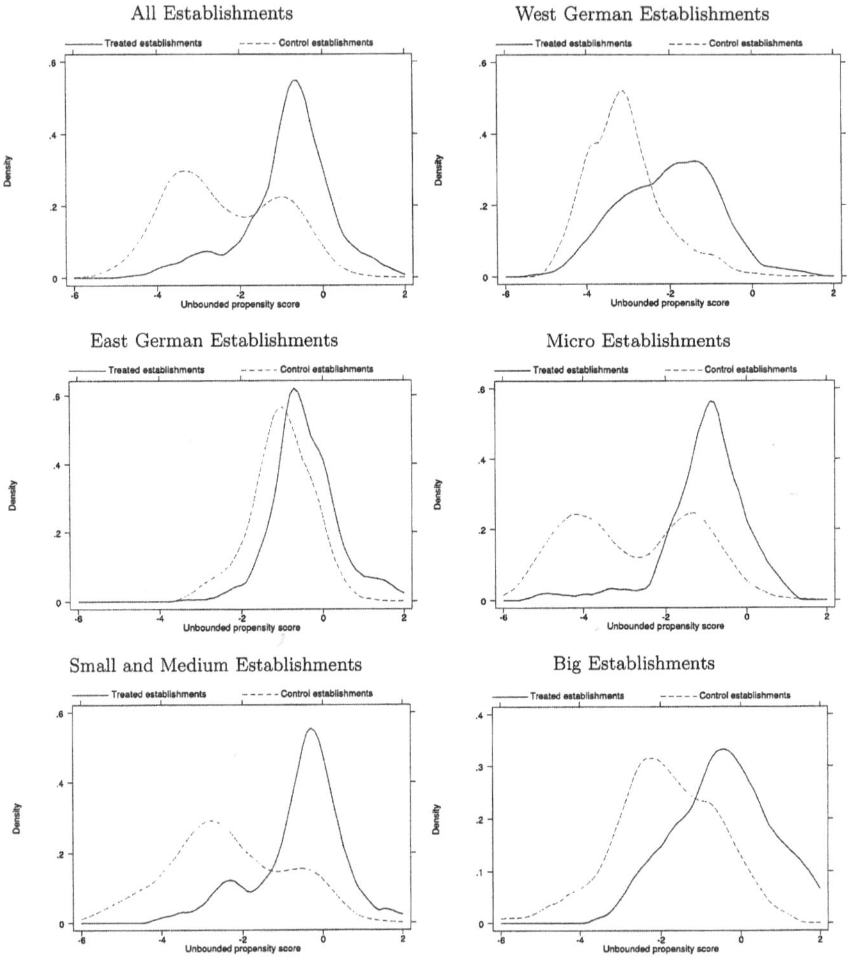

Appendix C

Additional Material to Chapter 4

C.1 Cluster and Correlation Analysis

A major problem of every cluster analysis is to answer the question how many clusters to form.[1] Thereby a trade-off between a low variance within each cluster and a preferable low number of clusters arises. In order to detect the optimal number of clusters we first of all computed a variety of dissimilarity measures depending on the number of clusters. Figure C.1 contains the root mean square standard deviation, R^2, the semi-partial R^2 and the cubic clustering criterion. The root mean square standard deviation should be as small as possible whereas the cubic clustering criterion should be maximized. Large jumps of the R^2 and the semi-partial R^2 give hints for an optimal number of clusters. Looking at figure C.1 one can identify that the optimal number of clusters lies between two and three.

Having identified the optimal number of clusters we conducted a non-hierarchical clustering with two and three clusters, respectively. Since the number of observations in the third cluster was rather small, namely 2,609 establishments for $Cluster_1$, 3,325 for $Cluster_2$ and only 248 for $Cluster_3$ we repeated the cluster analysis with only two clusters using an Euclidean distance measure and the standard Ward's method.

C.2 Some Speculation about Endogeneity

Although the estimation results were very robust regarding different models and specifications, one drawback which has to be taken into account is due to a possible endogeneity bias. In our econometric models we implicitly assumed that the causality runs from flexible workplace and HRM-practices to innovations. There

[1] For a discussion on cluster analysis see e.g. Sharma (1996).

Figure C.1: Detecting the Optimal Number of Clusters

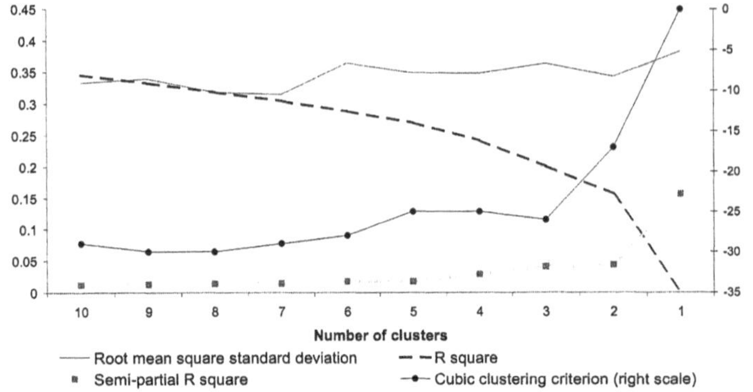

might be, however, also observable as well as unobservable covariates which have a common impact on both decisions thus leading to biased estimates. Endogeneity is simple to control for – if there were good instruments available. However, most of the variables referring to the outcome equation have also an impact on the decision whether to introduce organizational changes and vice versa.

To get an idea about the likely magnitude of this potential bias, we refer to chapter two where we have undertaken a small Monte Carlo simulation. The parameters used there are similar to the figures obtained from the trivariate Probit models. Figure 2.5 on page 51 contains the relative deviations of the estimated from the 'true' endogenous dummy parameters given different correlation coefficients between the outcome and treatment equation.

In the following we have estimated two Probit regressions with three different innovation dummies (Improvement, Incremental and Radical) and four different indicators reflecting flexible workplace and HRM-practices (HRM_2, HRM_3, HRM_4, $Cluster_2$), respectively. Thereby the same set of exogenous variables was used as before. Table C.1 contains for the so obtained residuals $\epsilon = y^* - x\beta$ the partial correlation coefficients. On average, the correlation between HRM_2 (HRM_3) and the innovation indicators is -0.46 (0.53). For HRM_4 and $Cluster_2$, the average $\hat{\rho}$ amounts to 0.72 and 0.69, respectively.

If the *dgp* assumed for the Monte Carlo study applies also to our application, these figures can be interpreted in the following manner: The impact of the HRM_2 dummy is underestimated whereas the impact of the remaining dummies on the outcome equation is overestimated. The corresponding relative deviations range from about -75% for HRM_2 to 100% and 150% for HRM_3 and $HRM_4/Cluster_2$. Although these results have to be treated with caution, they might give us an idea about the direction and likely magnitude of the endogeneity bias.

Table C.1: Estimated Partial Correlations

$\hat{\rho}$	HRM_2	HRM_2	HRM_2	$Cluster_2$
Improvement	-0.55	0.52	0.83	0.80
	(0.00)	(0.00)	(0.00)	(0.00)
Incremental	-0.38	0.56	0.64	0.62
	(0.00)	(0.00)	(0.00)	(0.00)
Radical	-0.45	0.51	0.68	0.64
	(0.00)	(0.00)	(0.00)	(0.00)

Note: p-values in parentheses.

C.3 Tables and Figures

Table C.2: Share of Establishments with Flexible Workplace and HRM Practices

	Before 1993	1995	1998	2000
		Germany		
Flexibility	n.a. (n.a.)	n.a. (n.a.)	22.41% (n.a.)	18.01% (8.28%)
Teamwork	n.a. (n.a.)	n.a. (n.a.)	15.58% (n.a.)	10.65% (4.99%)
Responsibility	n.a. (n.a.)	n.a. (n.a.)	13.38% (n.a.)	9.43% (5.57%)
		West Germany		
Flexibility	9.29% (9.29%)	31.95% (30.05%)	25.89% (18.24%)	22.04% (9.92%)
Teamwork	6.48% (6.48%)	19.55% (18.29%)	17.70% (12.17%)	12.87% (5.75%)
Responsibility	5.13% (5.13%)	14.32% (13.97%)	14.22% (12.17%)	10.82% (5.87%)
		East Germany		
Flexibility	n.a. (n.a.)	n.a. (n.a.)	19.24% (n.a.)	12.14% (6.91%)
Teamwork	n.a. (n.a.)	n.a. (n.a.)	13.64% (n.a.)	7.42% (4.32%)
Responsibility	n.a. (n.a.)	n.a. (n.a.)	12.62% (n.a.)	7.41% (5.31%)

Note: The first figure in the cells reflects the share of establishments including those which have already introduced organizational changes in the past while the figures in parentheses contain the share of establishments excluding those which have already introduced such an change in the last two/three years.

Table C.3: Share of Establishments with Flexible Workplace and HRM Practices (Cont.)

	Before 1993	1995	1998	2000
	Germany (≤ 5 Employees)			
Flexibility	n.a. (n.a.)	n.a. (n.a.)	4.10% (n.a.)	3.49% (1.32%)
Teamwork	n.a. (n.a.)	n.a. (n.a.)	2.14% (n.a.)	0.89% (0.56%)
Responsibility	n.a. (n.a.)	n.a. (n.a.)	1.05% (n.a.)	0.75% (0.38%)
	West Germany (≤ 5 Employees)			
Flexibility	0.84% (1.17%)	6.88% (6.11%)	3.51% (2.54%)	3.96% (0.83%)
Teamwork	6.48% (0.84%)	1.51% (1.52%)	1.72% (1.08%)	1.04% (0.45%)
Responsibility	0.67% (0.67%)	1.51% (1.35%)	0.86% (0.88%)	0.90% (0.45%)
	East Germany (≤ 5 Employees)			
Flexibility	n.a. (n.a.)	n.a. (n.a.)	4.60% (n.a.)	2.98% (1.77%)
Teamwork	n.a. (n.a.)	n.a. (n.a.)	2.51% (n.a.)	0.73% (0.66%)
Responsibility	n.a. (n.a.)	n.a. (n.a.)	1.22% (n.a.)	0.59% (0.32%)
	Germany (5-500 Employees)			
Flexibility	n.a. (n.a.)	n.a. (n.a.)	25.42% (n.a.)	22.02% (11.10%)
Teamwork	n.a. (n.a.)	n.a. (n.a.)	16.53% (n.a.)	12.36% (6.42%)
Responsibility	n.a. (n.a.)	n.a. (n.a.)	15.22% (n.a.)	10.95% (6.61%)
	West Germany (5-500 Employees)			
Flexibility	9.22% (9.22%)	29.70% (27.99%)	27.25% (19.09%)	25.79% (13.49%)
Teamwork	5.81% (5.81%)	15.79% (14.94%)	16.55% (11.60%)	14.11% (7.49%)
Responsibility	3.33% (3.33%)	11.70% (11.65%)	14.06% (10.77%)	11.71% (6.38%)
	East Germany (5-500 Employees)			
Flexibility	n.a. (n.a.)	n.a. (n.a.)	23.97% (n.a.)	16.21% (9.30%)
Teamwork	n.a. (n.a.)	n.a. (n.a.)	16.51% (n.a.)	9.66% (5.56%)
Responsibility	n.a. (n.a.)	n.a. (n.a.)	16.14% (n.a.)	9.76% (6.80%)
	Germany (> 500 Employees)			
Flexibility	n.a. (n.a.)	n.a. (n.a.)	47.36% (n.a.)	35.01% (19.32%)
Teamwork	n.a. (n.a.)	n.a. (n.a.)	40.05% (n.a.)	29.86% (13.19%)
Responsibility	n.a. (n.a.)	n.a. (n.a.)	31.08% (n.a.)	26.35% (17.61%)
	West Germany (> 500 Employees)			
Flexibility	14.09% (14.09%)	51.68% (51.00%)	54.76% (41.95%)	40.53% (25.75%)
Teamwork	11.31% (11.31%)	38.83% (37.51%)	45.13% (29.65%)	32.70% (14.52%)
Responsibility	11.98% (11.98%)	27.90% (28.21%)	34.57% (29.73%)	28.36% (18.59%)
	East Germany (> 500 Employees)			
Flexibility	n.a. (n.a.)	n.a. (n.a.)	4.60% (n.a.)	2.98% (1.77%)
Teamwork	n.a. (n.a.)	n.a. (n.a.)	29.34% (n.a.)	21.49% (11.44%)
Responsibility	n.a. (n.a.)	n.a. (n.a.)	23.72% (n.a.)	20.92% (16.08%)

Note: The first figure in the cells reflects the share of establishments including those which have already introduced organizational changes in the past while the figures in parentheses contain the share of establishments excluding those which have already introduced such an change in the last two/three years.

Table C.4: Share of Establishments with Flexible Workplace and HRM Practices (Cont.)

	Before 1993	1995	1998	2000
		Germany (Industry)		
Flexibility	n.a. (n.a.)	n.a. (n.a.)	31.04% (n.a.)	22.61% (12.85%)
Teamwork	n.a. (n.a.)	n.a. (n.a.)	23.65% (n.a.)	16.05% (7.09%)
Responsibility	n.a. (n.a.)	n.a. (n.a.)	18.39% (n.a.)	10.55% (7.16%)
		West Germany (Industry)		
Flexibility	9.80% (9.8%)	39.70% (38.78%)	36.37 (22.90%)	28.49% (18.06%)
Teamwork	7.47% (7.47%)	30.48% (29.10%)	28.99 (17.54%)	20.58% (8.67%)
Responsibility	12.41% (12.41%)	19.30% (18.97%)	20.58 (17.86%)	12.72% (9.35%)
		East Germany (Industry)		
Flexibility	n.a. (n.a.)	n.a. (n.a.)	24.44% (n.a.)	14.47% (7.67%)
Teamwork	n.a. (n.a.)	n.a. (n.a.)	17.04% (n.a.)	9.78% (5.52%)
Responsibility	n.a. (n.a.)	n.a. (n.a.)	15.68% (n.a.)	7.54% (4.71%)
		Germany (Service sector)		
Flexibility	n.a. (n.a.)	n.a. (n.a.)	24.46% (n.a.)	14.86% (9.16%)
Teamwork	n.a. (n.a.)	n.a. (n.a.)	16.87% (n.a.)	8.15% (4.79%)
Responsibility	n.a. (n.a.)	n.a. (n.a.)	13.27% (n.a.)	8.74% (6.09%)
		West Germany (Service sector)		
Flexibility	10.02% (10.02%)	28.63% (26.80%)	26.35% (18.50%)	17.40% (10.80%)
Teamwork	6.74% (6.74%)	14.18% (13.31%)	18.12% (12.78%)	9.24% (5.60%)
Responsibility	2.13% (2.13%)	12.32% (12.32%)	13.30% (12.21%)	9.47% (6.23%)
		East Germany (Service sector)		
Flexibility	n.a. (n.a.)	n.a. (n.a.)	22.15% (n.a.)	10.88% (7.33%)
Teamwork	n.a. (n.a.)	n.a. (n.a.)	15.35% (n.a.)	6.46% (3.88%)
Responsibility	n.a. (n.a.)	n.a. (n.a.)	13.21% (n.a.)	7.61% (5.92%)
		Germany (Industry related services)		
Flexibility	n.a. (n.a.)	n.a. (n.a.)	27.76% (n.a.)	16.28% (10.77%)
Teamwork	n.a. (n.a.)	n.a. (n.a.)	19.26% (n.a.)	8.61% (5.58%)
Responsibility	n.a. (n.a.)	n.a. (n.a.)	16.43% (n.a.)	8.61% (8.12%)
		West Germany (Industry related services)		
Flexibility	12.84% (12.84%)	34.90% (32.78%)	29.68% (18.81%)	18.44% (12.38%)
Teamwork	7.96% (7.96%)	15.55% (14.73%)	19.39% (15.07%)	9.74% (5.87%)
Responsibility	0.90% (0.9%)	15.73% (15.51%)	17.54% (17.97%)	9.94% (7.73%)
		East Germany (Industry related services)		
Flexibility	n.a. (n.a.)	n.a. (n.a.)	25.32% (n.a.)	12.21% (8.96%)
Teamwork	n.a. (n.a.)	n.a. (n.a.)	19.10% (n.a.)	6.49% (5.25%)
Responsibility	n.a. (n.a.)	n.a. (n.a.)	15.02% (n.a.)	6.10% (8.57%)

Note: The first figure in the cells reflects the share of establishments including those which have already introduced organizational changes in the past while the figures in parentheses contain the share of establishments excluding those which have already introduced such an change in the last two/three years.

Table C.5: Pearson Correlation Coefficient between Different Flexible Workplace and HRM Practices

	w_1	w_2	w_3	w_5	h_1	h_2	h_3	h_4	h_5	h_6	h_7	h_8	h_9	h_{10}	h_{11}	h_{12}	h_{13}	h_{14}	h_{15}
w_1	1.00	0.41	0.32	0.23	0.24	0.27	0.23	0.25	0.03	0.09	0.03	0.21	0.25	0.12	0.23	-0.01	-0.09	0.06	$0.13^{\dagger}$
w_2	0.41	1.00	0.28	0.20	0.21	0.24	0.21	0.24	0.05	0.08	0.03	0.20	0.20	0.10	0.21	-0.01	-0.10	0.06	$0.12^{\dagger}$
w_3	0.32	0.28	1.00	0.17	0.22	0.20	0.19	0.19	0.04	0.10	0.03	0.17	0.18	0.03	0.17	-0.01	-0.09	0.03	$0.09^{\dagger}$
w_4	0.23	0.20	0.17	1.00	0.24	0.24	0.16	0.24	0.04	0.12	0.03	0.24	0.25	0.13	0.22	-0.01	-0.10	0.03	$0.17^{\dagger}$
h_1	0.24	0.21	0.22	0.24	1.00	0.49	0.41	0.39	-0.03	0.18	0.01	0.49	0.22	-0.02	0.35	-0.02	-0.04	0.03	$0.15^{\dagger}$
h_2	0.27	0.24	0.20	0.24	0.49	1.00	0.42	0.41	-0.04	0.18	0.05	0.48	0.22	-0.03	0.35	-0.02	-0.07	0.03	$0.14^{\dagger}$
h_3	0.23	0.21	0.19	0.16	0.41	0.42	1.00	0.43	-0.03	0.17	0.02	0.41	0.16	-0.10	0.36	-0.02	-0.05	0.02	$0.17^{\dagger}$
h_4	0.25	0.24	0.19	0.24	0.39	0.41	0.43	1.00	0.00	0.14	0.03	0.37	0.22	0.00	0.30	-0.02	-0.09	0.03	$0.16^{\dagger}$
h_5	0.03	0.05	0.04	0.04	-0.03	-0.04	-0.03	0.00	1.00	0.04	0.03	-0.03	0.05	0.10	-0.02	0.00	-0.08	0.00	$-0.03^{\dagger}$
h_6	0.09	0.08	0.10	0.12	0.18	0.18	0.17	0.14	0.04	1.00	-0.01	0.26	0.04	-0.06	0.21	-0.01	-0.05	0.01	$0.06^{\dagger}$
h_7	0.03	0.03	0.03	0.03	0.01	0.05	0.02	0.03	0.03	-0.01	1.00	0.02	0.01	0.03	$0.02^{\dagger}$	0.00	0.06	0.00	$0.01^{\dagger}$
h_8	0.21	0.20	0.17	0.24	0.49	0.48	0.41	0.37	-0.03	0.26	0.02	1.00	0.18	-0.06	0.46	-0.03	-0.13	0.03	$0.12^{\dagger}$
h_9	0.25	0.20	0.18	0.25	0.22	0.22	0.16	0.22	0.05	0.04	0.01	0.18	1.00	0.22	0.17	-0.01	-0.10	0.05	$0.13^{\dagger}$
h_{10}	0.12	0.10	0.03	0.13	-0.02	-0.03	-0.10	0.00	0.10	-0.06	0.03	-0.06	0.22	1.00	$0.01^{\dagger}$	-0.03	-0.16	0.12	$0.02^{\dagger}$
h_{11}	0.23	0.21	0.17	0.22	0.35	0.35	0.36	0.30	-0.02	0.21	$0.02^{\dagger}$	0.46	0.17	$0.01^{\dagger}$	1.00	0.00	-0.09	0.04	0.46
h_{12}	-0.01	-0.01	-0.01	-0.01	-0.02	-0.02	-0.02	-0.02	0.00	-0.01	0.00	-0.03	-0.01	-0.03	0.00	1.00	-0.01	0.00	$0.01^{\dagger}$
h_{13}	-0.09	-0.10	-0.09	-0.10	-0.04	-0.07	-0.05	-0.09	-0.08	-0.05	0.06	-0.13	-0.10	-0.16	-0.09	-0.01	1.00	-0.08	$-0.01^{\dagger}$
h_{14}	0.06	0.06	0.03	0.03	0.03	0.03	0.02	0.03	0.00	0.01	0.00	0.03	0.05	0.12	0.04	0.00	-0.08	1.00	$0.01^{\dagger}$
h_{15}	$0.13^{\dagger}$	$0.12^{\dagger}$	$0.09^{\dagger}$	$0.17^{\dagger}$	$0.15^{\dagger}$	$0.14^{\dagger}$	$0.17^{\dagger}$	$0.16^{\dagger}$	$-0.03^{\dagger}$	$0.06^{\dagger}$	$0.01^{\dagger}$	$0.12^{\dagger}$	$0.13^{\dagger}$	$0.02^{\dagger}$	0.46	$0.01^{\dagger}$	$-0.01^{\dagger}$	$0.01^{\dagger}$	1.00

Notes: w_i refers to the i-th flexible workplace practice listed in table 4.4 whereas h_i indicates the i-th flexible HRM practice.
All correlation except those marked with a † were significant at least on a 5%-level.

Appendix D

Additional Material to Chapter 5

D.1 Tables

Table D.1: Means of the Variables Used for the Estimation (Standard Errors in Parentheses)

Variable	Description	Overall	$Y = 0$
$STATE_{98}$	State of technology (1: Up-to-data, ..., 5: Out-of-date)	2.10 (0.79)	2.12 (0.81)
ICT_{97}	Investment in ICT (yes/no)	0.53 (0.50)	0.37 (0.48)
$HIGH_{98}$	Share of employees with an university degree	0.05 (0.11)	0.03 (0.11)
$YOUNG_{98}$	Share of younger employees between 20 and 40	0.47 (0.25)	0.48 (0.30)
$EMPL_{98}$	Total number of employees	303 (1,170)	92 (357)
$COMP_{98}$	Competition pressure (1 = None, ..., 4 = High)	3.51 (0.78)	3.40 (0.84)
UP_{98}	Business volume expected to increase in 1998	0.29 (0.45)	0.23 (0.42)
EXP_{98}	Export share in 1998	0.07 (0.17)	0.03 (0.12)
EAST	Establishment located in East Germany	0.51 (–)	0.54 (–)

Table D.2: Relative Importance of the Lindbeck/Snower Factors

Factor	Standard NegBin I Coefficient	Standard NegBin I t-Value	Hurdle NegBin I Coefficient	Hurdle NegBin I t-Value	Truncated NegBin I Coefficient	Truncated NegBin I t-Value
	West German establishment – Service Sector					
Versatile Capital	0.07	1.46	0.08	1.53	0.19	1.49
Investment ICT	0.35	6.56	0.31	3.96	0.28	3.32
Human Capital	0.20	1.84	0.17	1.44	0.28	1.86
Worker Tastes	0.38	6.40	0.45	3.33	0.25	2.80
Total ΔY	2.64	1.52	0.47	2.16	3.09	1.55
	East German establishment – Investment Goods Industry					
Versatile Capital	0.08	1.52	0.06	1.38	0.19	1.48
Investment ICT	0.35	6.22	0.30	6.14	0.28	3.41
Human Capital	0.20	1.85	0.15	1.41	0.29	1.88
Worker Tastes	0.37	6.41	0.49	6.44	0.25	2.95
Total ΔY	2.58	6.24	0.62	6.01	1.36	2.73
	East German establishment – Service Sector					
Versatile Capital	0.08	1.47	0.07	1.30	0.19	1.48
Investment ICT	0.35	6.09	0.30	3.94	0.28	3.38
Human Capital	0.19	1.79	0.17	1.46	0.28	1.85
Worker Tastes	0.38	6.45	0.46	3.59	0.25	2.84
Total ΔY	2.39	1.44	0.46	1.99	2.54	1.59

Appendix E

Additional Material to Chapter 6

E.1 Variables Used for the Estimation

Table E.1: Variables Used for the Employment Shares in 1998[a]

Variable	Description
INV_{97}/Y_{97}	Logarithm of Total Investments in 1997 per Business Volume (in DM)
DEV_{97}	Business Development in 1997 (1 = Very good, ..., 5 = Insufficient)
ICT_{97}	Investment in Information and Communication Technology in 1997 (Yes/No)
$COMP$	Competition Pressure (1 = No Pressure, ..., 4 = High Pressure)
$STATE_{98}$	State of the Technology in 1998 (1 = Up-to-Date, ..., 5 = Out-of-Date)
$INNO$	Product Improvement, Incremental or Radical Innovation (Yes/No)[b]
$ORGA$	Introduced Organizational Changes (Yes/No)[c]
	16 Industry Dummies with Reference Class Agriculture
	16 Regional Dummies with Reference Class Eastern Berlin

[a] Share of high skilled employees thereby of skilled workers and white Collar workers for qualified Tasks. Low Skilled employees are made up of un- and semiskilled blue collar workers and white collar workers and officials for simple tasks.
[b] Improvement of existing products and services, introduction of new products and services for the own establishment or for the market during 1996 and 1997.
[c] Reorganization of departments and functional areas, decentralization of responsibilities and decision-authority, introduction of group-Workplaces with own responsibilities and introduction of units with own costing and result calculation during 1996 and 1997.

Table E.2: Variables Used for the Innovation Decision[a]

Variable	Description
$HIGH_{96}$	Share of High Qualified White and Blue Collar Employees in 1996
$R\&D$	R&D Department (Yes/No)
MR	Market Research Department (Yes/No)
$ORGA$	Introduced Organizational Changes (Yes/No)[b]
$PROF$	Profit Participation Plans for Employees (Yes/No)
$COOP$	Engagement in R&D Cooperation with other Institutions
INV_{97}	Total Investments in 1997 (in DM)
ICT_{95}	Investment in Information and Communication Technology in 1995 (Yes/No)
$STATE_{96}$	State of the Technology in 1996 (1 = Up-to-Date, ..., 5 = Out-of-Date)
DEV_{97}	Business Development in 1997 (1 = Very good, ..., 5 = Insufficient)
$COMP$	Competition Pressure (1 = No Pressure, ..., 4 = High Pressure)
$ONEMAN$	One Man Business[c]
$PART$	Establishment Organized as a Business Partnership[c]
$LIMTD$	Private Limited Company[c]
$CAPT$	Capital Company[c]
$PUBL$	Public Corporation[c]
$SIZE_{50}$	Establishment with Less than 50 Employees
$SIZE_{500}$	Establishment with Less than 500 Employees
	16 Industry Dummies with Reference Class Agriculture
	16 Regional Dummies with Reference Class Eastern Berlin

[a] Improvement of existing products and services, introduction of new products and services for the own establishment or for the market during 1996 and 1997.
[b] Reorganization of departments and functional areas, decentralization of responsibilities and decision-authority, introduction of group-Workplaces with own responsibilities and introduction of units with own costing and result calculation during 1996 and 1997.
[c] Reference class consists of miscellaneous other legal forms, e.g. cooperatives and incorporated societies.

Table E.3: Variables Used for Introduced Organizational Changes[a]

Variable	Description
$HIGH_{96}$	Share of High Qualified White and Blue Collar Employees in 1996
$COMP$	Competition Pressure (1 = No Pressure, ..., 4 = High Pressure)
DEV_{96}	Business Development in 1996 (1 = Very good, ..., 5 = Insufficient)
DEV_{97}	Business Development in 1997 (1 = Very good, ..., 5 = Insufficient)
$INNO$	Product Improvement, Incremental or Radical Innovation (Yes/No)[b]
INV_{97}	Total Investments in 1997 (in DM)
ICT_{95}	Investment in Information and Communication Technology in 1995 (Yes/No)
ICT_{97}	Investment in Information and Communication Technology in 1997 (Yes/No)
$STATE_{96}$	State of the Technology in 1996 (1 = Up-to-Date, ..., 5 = Out-of-Date)
$UNION$	Workers/Staff Council (Yes/No)
$ONEMAN$	One Man Business[c]
$PART$	Establishment Organized as a Business Partnership[c]
$LIMTD$	Private Limited Company[c]
$CAPT$	Capital Company[c]
$PUBL$	Public Corporation[c]
$SIZE_{50}$	Establishment with Less than 50 Employees
$SIZE_{500}$	Establishment with Less than 500 Employees
	16 Industry Dummies with Reference Class Agriculture
	16 Country Dummies with Reference Class Eastern Berlin

[a] Reorganization of departments and functional areas, decentralization of responsibilities and decision-authority, introduction of group-Workplaces with own responsibilities and introduction of units with own costing and result calculation during 1996 and 1997.

[b] Improvement of existing products and services, introduction of new products and services for the own establishment or for the market during 1996 and 1997.

[c] Reference class consists of miscellaneous other legal forms, e.g. cooperatives and incorporated societies.

E.2 Tables and Figures

Table E.4: Estimation Results Type I Model: Share Equation for High Skilled Employees

	OLS		Amemiya		Nelson-Olson	
Variable	Parameter	t-Value	Parameter	t-Value	Parameter	t-Value
$Intercept$	75.2796	18.01	35.6447	2.60	83.3970	18.95
INV_{97}/Y_{97}	0.0023	0.01	0.2902	0.03	-0.4913	-1.37
DEV_{97}	0.8410	1.87	0.9076	0.10	0.5990	1.29
ICT_{97}	4.4372	4.12	1.9770	0.20	2.1220	1.72
$COMP$	-0.0024	-0.00	0.5815	0.06	-0.6359	-0.89
$STATE_{98}$	-1.2776	-2.02	-1.7081	-0.19	-0.8970	-1.34
$INNO$	3.1828	3.03	-4.5698	-0.49	3.4531	2.18
$ORGA$	2.2563	2.19	0.9922	0.29	1.5477	1.13
	F-Value		F-Value		F-Value	
Industry	19.08		2.40		17.30	
Region	18.38		3.27		19.62	
R^2_{adj}	0.1677		–		0.1820	

Table E.5: Estimation Results Type I Model: Share Equation for Low Skilled Employees

	OLS		Amemiya		Nelson-Olson	
Variable	Parameter	t-Value	Parameter	t-Value	Parameter	t-Value
$Intercept$	17.3708	4.16	56.8459	5.27	10.3330	2.33
INV_{97}/Y_{97}	-0.1142	-0.34	-0.4634	-0.06	0.2550	0.71
DEV	-0.7244	-1.61	-1.0794	-0.15	-0.6593	-1.41
ICT_{97}	-3.3870	-3.15	-1.3579	-0.18	-2.0079	-1.61
$COMP$	0.0514	0.08	-0.5794	-0.08	0.5239	0.73
$STATE_{98}$	1.6038	2.54	2.1076	0.29	1.4275	2.12
$INNO$	-2.8812	-2.75	5.3809	0.73	-2.3635	-1.48
$ORGA$	-2.2677	-2.21	-1.8820	-0.70	-1.0399	-0.76
	F-Value		F-Value		F-Value	
Industry	21.74		4.63		19.71	
Region	21.90		4.53		22.24	
R^2_{adj}	0.1893		–		0.1949	

Table E.6: Estimation Results Type I Model: Innovation Decision

	ML		Amemiya		Nelson-Olson	
Variable	Parameter	t-Value	Parameter	t-Value	Parameter	t-Value
Intercept	-1.2680	-5.06	-0.3896	-29.9237	-0.6247	-2.03
$HIGH_{96}$	0.0015	1.92	0.0003	0.1100	0.0003	0.25
R&D	0.6502	2.98	0.8459	65.4328	0.8298	3.08
MR	0.3591	7.60	0.1739	34.3940	0.1756	2.12
PROF	0.1891	2.60	0.0335	4.7506	0.0269	0.23
COOP	0.3069	1.35	-0.0686	-3.4896	-0.0374	-0.13
INV_{97}	1.19E-9	1.09	-4.50E-010	-1.58E-007	1.13E-10	0.09
ICT_{95}	0.2532	5.43	0.0374	9.6453	0.0492	0.66
$STATE_{96}$	-0.1036	-3.80	-0.0772	-26.9107	-0.1075	-3.19
DEV_{97}	-0.0192	-0.91	-0.0120	-4.1653	-0.0261	-0.94
COMP	0.1262	4.23	0.0638	22.1078	0.0557	1.42
$SIZE_{50}$	-0.2418	-2.42	-0.0373	-4.2024	0.0007	0.00
$SIZE_{500}$	-0.3097	-3.29	-0.1767	-32.7806	-0.1733	-1.38
ONEMAN	0.0433	0.36	0.1734	49.5953	0.1823	1.18
PART	0.1360	1.04	0.2256	73.8131	0.2233	1.40
LIMTD	0.2362	2.07	0.2919	97.4241	0.2869	2.05
CAPT	0.1659	1.17	0.2946	75.0272	0.2520	1.42
PUBL	0.0782	0.49	0.0999	22.3531	0.0748	0.34
ORGA	0.5967	12.17	0.4748	38.2437	0.4719	3.67
	LR-Value		F-Value		LR-Value	
Industry	137.2174		91.76		96.9473	
Region	24.7570		3.73		11.6201	
LR-Test	8,121.8093		–		5,273.5230	

Table E.7: Estimation Results Type I Model: Organizational Changes

	ML		Amemiya		Nelson-Olson	
Variable	Parameter	t-Value	Parameter	t-Value	Parameter	t-Value
Intercept	-0.5805	-2.25	-0.2500	-0.3285	-0.1635	-0.55
$HIGH_{96}$	0.0016	1.91	0.0004	0.0011	0.0007	0.72
$COMP$	0.1019	3.10	-0.0179	-0.0468	0.0278	0.71
DEV_{96}	0.0412	1.70	0.0460	0.1218	0.0520	1.84
DEV_{97}	0.0269	1.13	0.0423	0.1121	0.0441	1.59
INV_{97}	1.83E-9	1.27	1.57E-009	4.17E-009	1.57E-9	1.08
ICT_{95}	0.2424	4.97	0.1482	0.3755	0.1862	3.14
ICT_{97}	0.4402	8.77	0.2301	0.5632	0.3030	4.78
$UNION$	0.2404	3.76	0.1839	0.4849	0.2101	2.83
$SIZE_{50}$	-0.8874	-8.16	-0.4021	-0.8840	-0.4558	-3.61
$SIZE_{500}$	-0.5118	-5.22	-0.1394	-0.2997	-0.1962	-1.78
$ONEMAN$	-0.4567	-3.50	-0.2803	-0.7377	-0.2682	-1.80
$PART$	-0.3350	-2.41	-0.2603	-0.6624	-0.2457	-1.56
$LIMTD$	-0.0822	-0.67	-0.1659	-0.3728	-0.0927	-0.67
$CAPT$	0.23699	1.57	-0.0528	-0.1088	0.1020	0.59
$PUBL$	0.3815	1.79	0.1443	0.3438	0.2420	1.06
$INNO$	0.6360	13.25	0.7996	1.3982	0.5768	9.24
	LR-Value		F-Value		LR-Value	
Industry	12.1841		0.13		29.0652	
Region	23.7654		0.03		18.6007	
LR-Test	9,703.0617		–		6,314.8505	

Table E.8: Type II Model ($INNO \rightarrow ORGA$): High Skilled Employees

Variable	OLS Parameter	OLS t-Value	IV-White Parameter	IV-White t-Value
Intercept	75.2796	18.01	75.7018	17.41
INV_{97}/Y_{97}	0.0023	0.01	-0.2781	-0.77
DEV_{97}	0.8410	1.87	0.7518	1.62
ICT_{97}	4.4372	4.12	2.9235	2.28
$COMP$	-0.0024	-0.00	-1.1326	-1.40
$STATE_{98}$	-1.2776	-2.02	-0.8715	-1.34
$INNO$	3.1823	3.03	17.7980	3.36
$ORGA$	2.2563	2.19	1.2661	0.22
	F-Value		F-Value	
Industry	19.08		18.44	
Region	18.38		20.06	
R^2_{adj}	0.1677		0.1867	

Table E.9: Type II Model ($INNO \rightarrow ORGA$): Low Skilled Employees

Variable	OLS Parameter	OLS t-Value	IV-White Parameter	IV-White t-Value
Intercept	17.3708	4.16	15.2136	3.48
INV_{97}/Y_{97}	-0.1142	-0.34	0.1556	0.43
DEV	-0.7244	-1.61	-0.8512	-1.85
ICT_{97}	-3.3870	-3.15	-3.0346	-2.35
$COMP$	0.0514	0.08	0.9132	1.10
$STATE_{98}$	1.6038	2.54	1.3582	2.07
$INNO$	-2.8812	-2.75	-18.7135	-3.46
$ORGA$	-2.2677	-2.21	5.0413	0.88
	F-Value		F-Value	
Industry	21.74		21.71	
Region	21.90		23.06	
R^2_{adj}	0.1893		0.1998	

Table E.10: Type II Model ($INNO \rightarrow ORGA$): Innovation Decision[a]

	ML		IV-White	
Variable	Parameter	t-Value	Parameter	t-Value
Intercept	-1.0186	-4.14	-1.0186	-4.14
$HIGH_{96}$	0.0018	2.33	0.0018	2.33
R&D	0.5023	2.30	0.5023	2.30
MR	0.4450	9.65	0.4450	9.65
$PROF$	0.3010	4.20	0.3010	4.20
$COOP$	0.4859	2.14	0.4859	2.14
INV_{97}	1.35E-9	12.23	1.35E-9	12.23
ICT_{95}	0.3131	6.85	0.3131	6.85
$STATE_{96}$	-0.0992	-3.69	-0.0992	-3.69
DEV_{97}	-0.0121	-0.58	-0.0121	-0.58
$COMP$	0.1368	4.65	0.1368	4.65
$SIZE_{50}$	-0.4281	-4.39	-0.4281	-4.39
$SIZE_{500}$	-0.3851	-4.15	-0.3851	-4.15
$ONEMAN$	-0.0347	-0.29	-0.0347	-0.29
$PART$	0.0737	0.57	0.0737	0.57
$LIMTD$	0.2320	2.08	0.2320	2.08
$CAPT$	0.2135	1.52	0.2135	1.52
$PUBL$	0.1174	0.74	0.1174	0.74
$ORGA$	–	–	–	–
	LR-Value		LR-Value	
Industry	142.0695		142.0695	
Region	30.7014		30.7014	
LR-Test	3,624.4956		3,624.4956	

[a] Note that for this type II model, the innovation decision is the first equation to be estimated and therefore the estimation results in the columns ML and IV-White are identical. The same holds true for table E.15

Table E.11: Type II Model ($INNO \rightarrow ORGA$): Organizational Changes

Variable	ML Parameter	ML t-Value	IV-White Parameter	IV-White t-Value
Intercept	-0.5805	-2.25	-0.8670	-3.35
$HIGH_{96}$	0.0016	1.91	0.0003	0.34
$COMP$	0.1019	3.10	0.0314	0.93
DEV_{96}	0.0412	1.70	0.0425	1.80
DEV_{97}	0.0269	1.13	0.0353	1.49
INV_{97}	1.83E-9	1.27	1.27E-09	0.76
ICT_{95}	0.2424	4.97	0.0670	1.25
ICT_{97}	0.4402	8.77	0.4401	8.80
$UNION$	0.2404	3.76	0.2241	3.51
$SIZE_{50}$	-0.8874	-8.16	-0.5832	-5.06
$SIZE_{500}$	-0.5118	-5.22	-0.3204	-3.11
$ONEMAN$	-0.4567	-3.50	-0.3754	-2.87
$PART$	-0.3350	-2.41	-0.3463	-2.52
$LIMTD$	-0.0822	-0.67	-0.1814	-1.46
$CAPT$	0.2369	1.57	0.1616	1.07
$PUBL$	0.3815	1.79	0.2938	1.38
$INNO$	0.6360	13.25	1.8777	10.89
	LR-Value		LR-Value	
Industry	12.1841		35.4848	
Region	23.7654		19.4927	
LR-Test	9,703.0617		5,165.9397	

Table E.12: Type II Model ($ORGA \rightarrow INNO$): High Skilled Employees

Variable	OLS Parameter	OLS t-Value	IV-White Parameter	IV-White t-Value
Intercept	75.2796	18.01	74.5356	16.88
INV_{97}/Y_{97}	0.0023	0.01	-0.1457	-0.40
DEV_{97}	0.8410	1.87	0.5416	1.18
ICT_{97}	4.4372	4.12	0.3758	0.28
$COMP$	-0.0024	-0.00	-1.1408	-1.40
$STATE_{98}$	-1.2776	-2.02	-1.0265	-1.57
$INNO$	3.1828	3.03	21.2834	4.19
$ORGA$	2.2563	2.19	-0.1873	-0.04
	F-Value		F-Value	
Industry	19.08		18.88	
Region	18.38		19.83	
R^2_{adj}	0.1677		0.1867	

Table E.13: Type II Model ($ORGA \rightarrow INNO$): Low Skilled Employees

Variable	OLS Parameter	OLS t-Value	IV-White Parameter	IV-White t-Value
Intercept	17.3708	4.16	17.2763	3.90
INV_{97}/Y_{97}	-0.1142	-0.34	0.1296	0.04
DEV	-0.7244	-1.61	-0.6591	-1.45
ICT_{97}	-3.3870	-3.15	-1.0882	-0.80
$COMP$	0.0514	0.08	0.8496	1.02
$STATE_{98}$	1.6038	2.54	1.4587	2.22
$INNO$	-2.8812	-2.75	-19.5096	-3.95
$ORGA$	-2.2677	-2.21	5.2894	1.26
	F-Value		F-Value	
Industry	21.74		21.41	
Region	21.90		22.53	
R^2_{adj}	0.1893		0.1977	

Table E.14: Type II Model ($ORGA \rightarrow INNO$): Innovation Decision

	ML		IV-White	
Variable	Parameter	t-Value	Parameter	t-Value
Intercept	-1.2680	-5.06	-1.7469	-5.98
$HIGH_{96}$	0.0015	1.92	0.0003	0.38
$R\&D$	0.6502	2.98	0.5489	2.44
MR	0.3591	7.60	0.3910	8.19
$PROF$	0.1891	2.60	0.2742	3.72
$COOP$	0.3069	1.35	0.4068	1.74
INV_{97}	1.19E-9	1.09	1.08E-09	0.92
ICT_{95}	0.2532	5.43	0.0847	1.46
$STATE_{96}$	-0.1036	-3.80	-0.1064	-3.86
DEV_{97}	-0.0192	-0.91	-0.0328	-1.52
$COMP$	0.1262	4.23	0.0557	1.66
$SIZE_{50}$	-0.2418	-2.42	0.1928	1.34
$SIZE_{500}$	-0.3097	-3.29	-0.1050	-0.97
$ONEMAN$	0.0433	0.36	0.2014	1.43
$PART$	0.1360	1.04	0.2045	1.41
$LIMTD$	0.2362	2.07	0.2336	1.82
$CAPT$	0.1659	1.17	0.0816	0.52
$PUBL$	0.0782	0.49	-0.0107	-0.05
$ORGA$	0.5967	12.17	1.6293	6.36
	LR-Value		LR-Value	
Industry	137.2174		23.3733	
Region	24.7570		29.3530	
LR-Test	8,121.8093		5,020.4088	

Table E.15: Type II Model ($ORGA \rightarrow INNO$): Organizational Changes

Variable	ML		IV-White	
	Parameter	t-Value	Parameter	t-Value
Intercept	-0.5295	-2.09	-0.5295	-2.09
$HIGH_{96}$	0.0021	2.56	0.0021	2.56
$COMP$	0.1287	4.00	0.1287	4.00
DEV_{96}	0.0403	1.70	0.0403	1.70
DEV_{97}	0.0225	0.97	0.0225	0.97
INV_{97}	1.98E-09	1.42	1.98E-09	1.42
ICT_{95}	0.3000	6.31	0.3000	6.31
ICT_{97}	0.5059	10.36	0.5059	10.36
$UNION$	0.2649	4.23	0.2649	4.23
$SIZE_{50}$	-0.9361	-8.88	-0.9361	-8.88
$SIZE_{500}$	-0.5665	-5.97	-0.5665	-5.97
$ONEMAN$	-0.4322	-3.40	-0.4322	-3.40
$PART$	-0.2828	-2.09	-0.2828	-2.09
$LIMTD$	0.0002	0.00	0.0002	0.00
$CAPT$	0.3173	2.16	0.3173	2.16
$PUBL$	0.3752	1.84	0.3752	1.84
$INNO$	-	-	-	-
	LR-Value		LR-Value	
Industry	23.3733		23.3733	
Region	29.3530		29.3530	
LR-Test	5,020.4088		5,020.4088	

Table E.16: Pearson Correlation Coefficient between Different Forms of Innovation

	Improvement	Incremental	Radical
Improvement	1.00 (–)	0.44 (<.001)	0.29 (<.001)
Incremental		1.00 (–)	0.24 (<.001)
Radical			1.00 (–)

Note: Significance levels in parentheses.

Table E.17: Pearson Correlation Coefficient between Different Organizational Changes

	Reorganization	Decentralization	Responsibility	Own Costing
Reorganization	1.00 (–)	0.44 (<.001)	0.36 (<.001)	0.35 (<.001)
Decentralization		1.00 (–)	0.42 (<.001)	0.34 (<.001)
Responsibility			1.00 (–)	0.29 (<.001)
Own Costing				1.00 (–)

Note: Significance levels in parentheses.

Table E.18: Percentage of Establishments Falling in Each Category

		Organizational Change	
		Yes	No
Innovation Decision	Yes	38%	15%
	No	15%	31%

Figure E.1: Affiliation to Industrial Sectors of Establishments that Participated in 1998 and 1996 at the IAB Panel

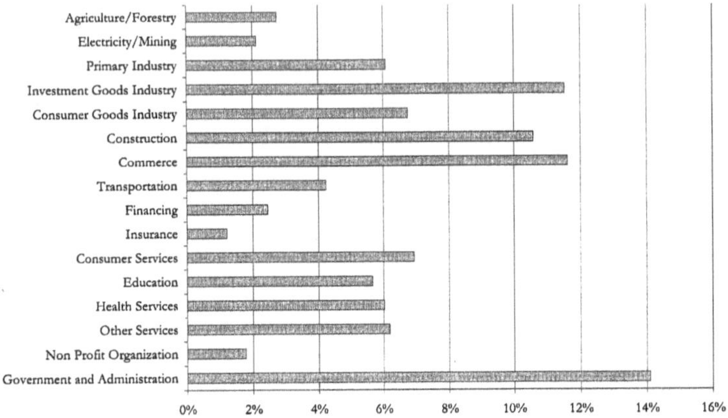

Figure E.2: Size Class of Establishments that Participated in 1998 and 1996 at the IAB Panel

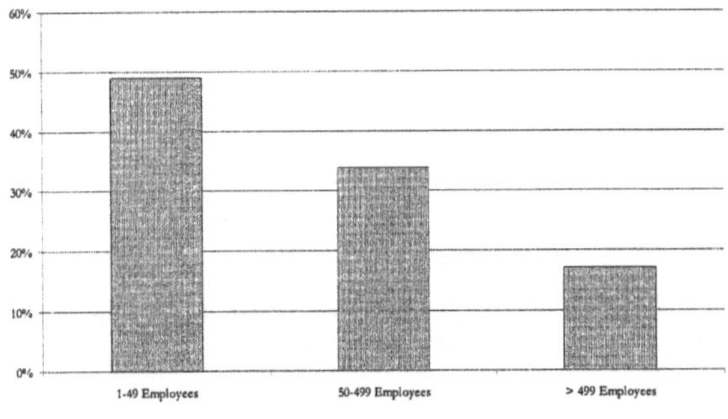

Figure E.3: Innovation Decision with Regard to the Size Class of the Establishments

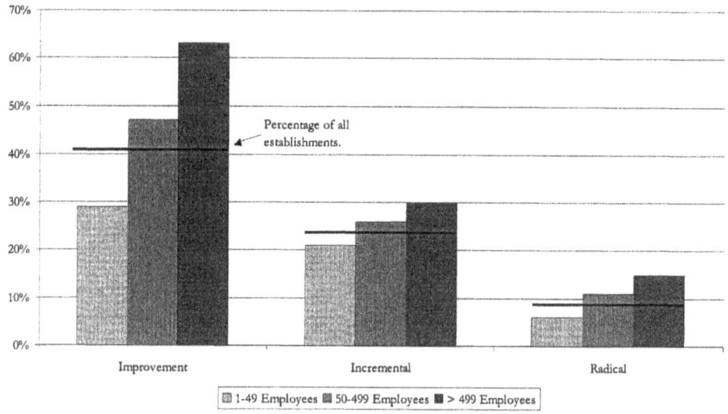

Figure E.4: Innovation Decision for West- and East German Establishments

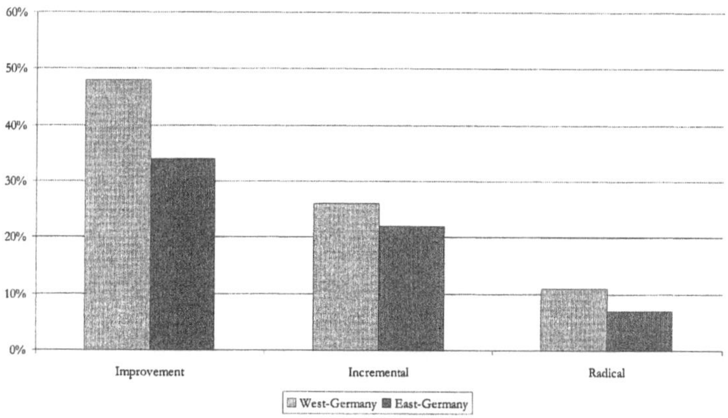

Figure E.5: Improvement of Existing Products and Services with Regard to the Industrial Affiliation of the Establishments

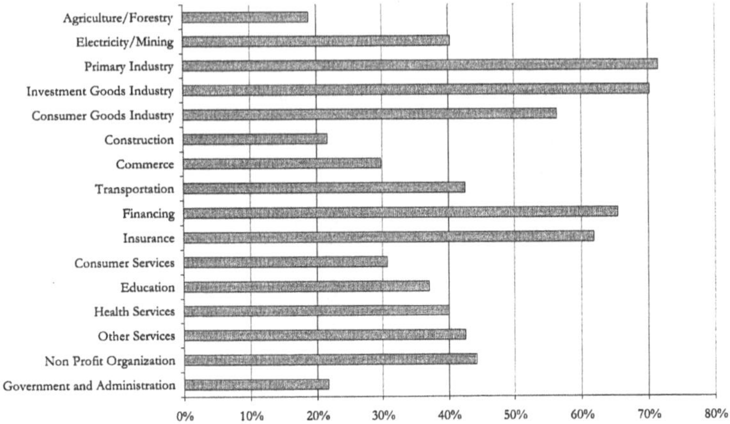

Figure E.6: Introduction of Products and Services that are New for the Own Establishment with Regard to the Industrial Affiliation of the Establishments

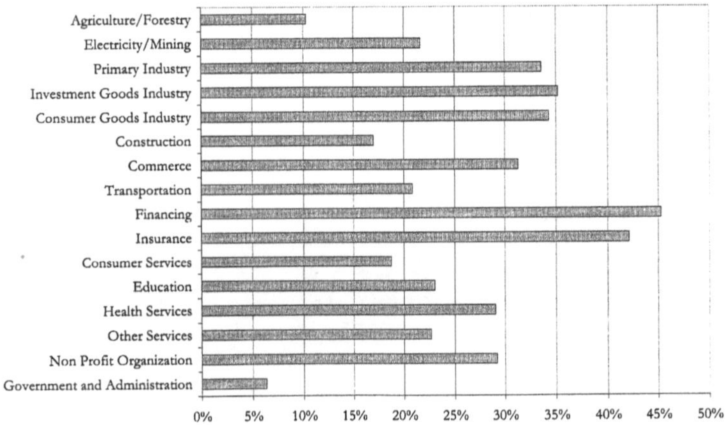

Figure E.7: Introduction of Products and Services that are New for the Market with Regard to the Industrial Affiliation of the Establishments

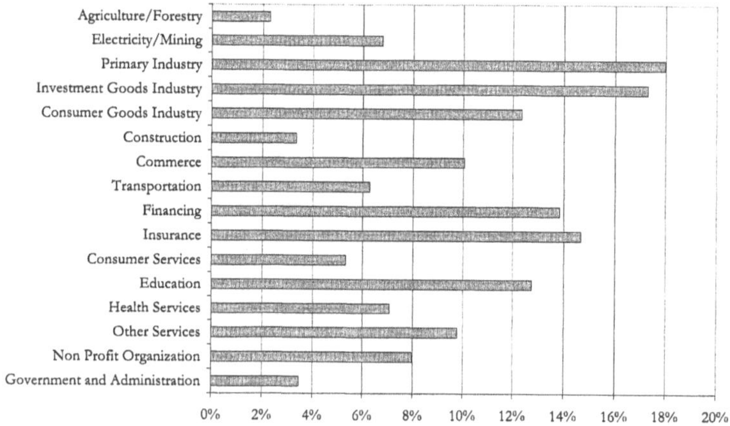

Figure E.8: Organizational Changes with Regard to the Size Class of the Establishments

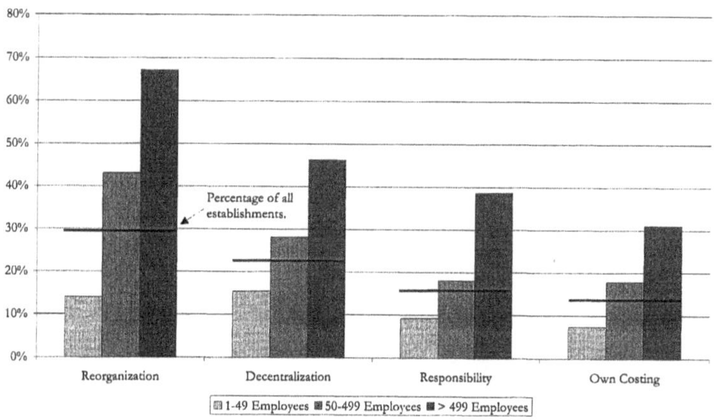

Figure E.9: Organizational Changes for West- and East German Establishments

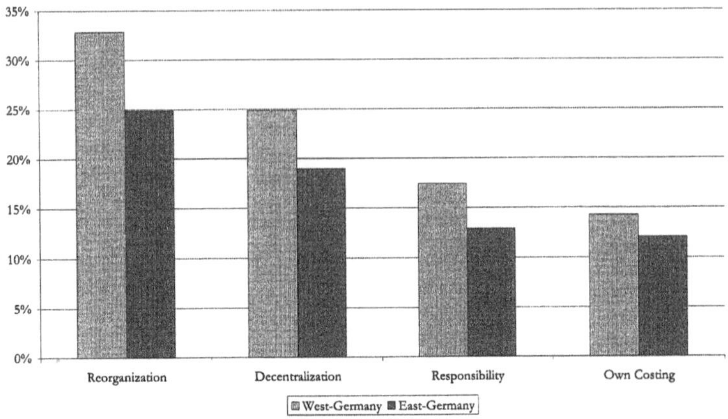

Figure E.10: Reorganization of Departments and Functional Areas with Regard to the Industrial Affiliation of the Establishments

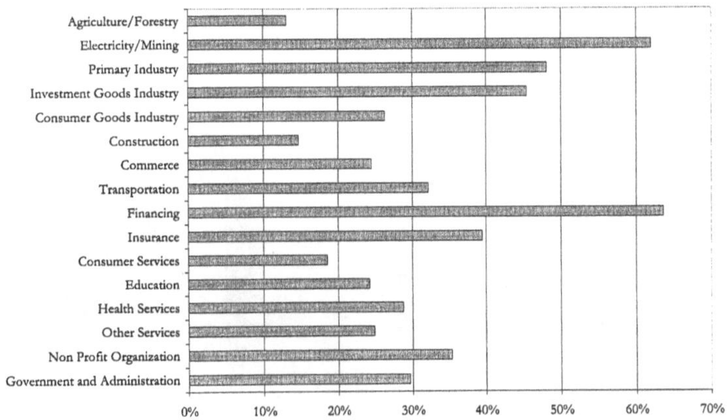

Figure E.11: Decentralization of Responsibilities and Decision-Authority with Regard to the Industrial Affiliation of the Establishments

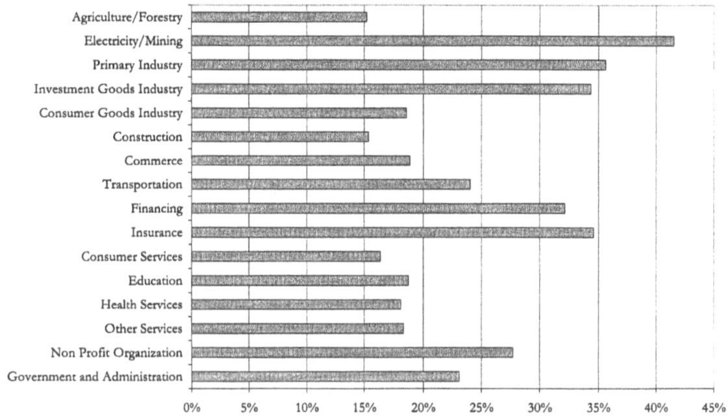

Figure E.12: Introduction of Group-Workplaces with Own Responsibilities with Regard to the Industrial Affiliation of the Establishments

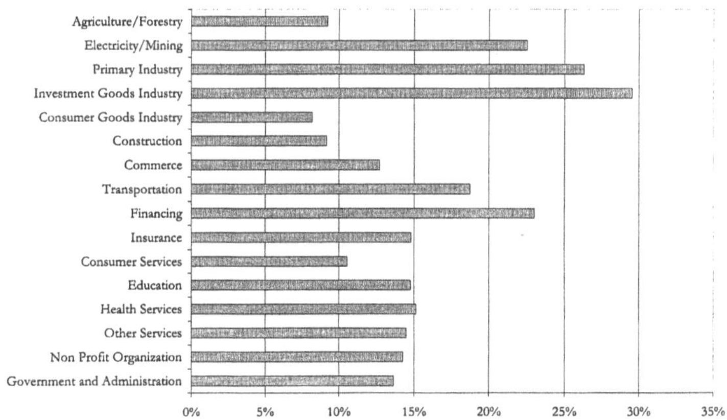

Figure E.13: Introduction of Units with Own Costing and Result Calculation with Regard to the Industrial Affiliation of the Establishments

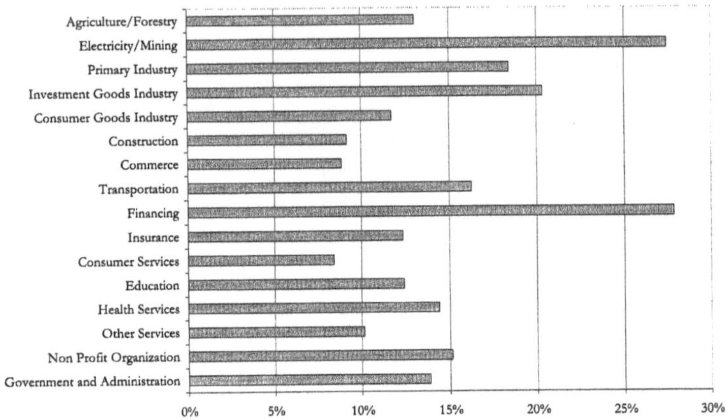

Appendix F

Additional Material to Chapter 7

F.1 Tables

Table F.1: SURE Estimation Results for West Germany

	High Skilled		Medium Skilled		Low Skilled	
	Parameter	SE	Parameter	SE	Parameter	SE
Intercept	-0.781	0.222	0.79	0.407	1.67	0.245
w_M/w_H	1.18	0.243	1.18	0.243	-0.099	0.138
w_L/w_H	-0.099	0.138	-0.508	0.197	-0.508	0.197
INV_{97}	-0.000195	0.000245	-0.00177	0.000643	-0.00104	0.000427
INNO	0.0912	0.0395	-0.149	0.104	0.377	0.0687
ORGA	-0.029	0.034	-0.0538	0.0898	-0.397	0.0594
	LR-Value		LR-Value		LR-Value	
Industry	171.15		380.03		198.25	
LR-Test	176.07		471.48		262.78	

Table F.2: SURE Estimation Results for East Germany

	High Skilled		Medium Skilled		Low Skilled	
	Parameter	SE	Parameter	SE	Parameter	SE
Intercept	0.499	0.539	0.839	0.905	-0.799	0.374
w_M/w_H	0.386	0.584	0.386	0.584	-0.459	0.225
w_L/w_H	-0.459	0.225	1.24	0.295	1.24	0.295
INV_{97}	-1.94·10-9	9.75·10-10	-1.17·10-9	2.75·10-9	1.33·10-9	1.44·10-9
INNO	0.0008	0.0384	0.0469	0.108	-0.0249	0.0566
ORGA	0.0403	0.0668	-0.448	0.189	-0.144	0.0988
	LR-Value		LR-Value		LR-Value	
Industry	95.05		85.09		198.57	
LR-Test	133.41		108.93		297.14	

Table F.3: Own and Cross Price Elasticities for West Germany

	High Skilled	Medium Skilled	Low Skilled
High Skilled	-1.70	1.83	-0.13
	(0.46)	(0.47)	(0.19)
Medium Skilled	0.41	-0.28	-0.13
	(0.1)	(0.12)	(0.07)
Low Skilled	-0.08	-0.36	0.45
	(0.12)	(0.21)	(0.17)

Note: Standard errors in parentheses.

Table F.4: Own and Cross Price Elasticities for East Germany

	High Skilled	Medium Skilled	Low Skilled
High Skilled	-0.02	0.18	-0.16
	(0.32)	(0.3)	(0.1)
Medium Skilled	0.06	-0.19	0.13
	(0.1)	(0.12)	(0.05)
Low Skilled	-0.30	0.69	-0.39
	(0.21)	(0.31)	(0.21)

Note: Standard errors in parentheses.

Table F.5: Estimation Results for the Innovation Decision

	West Germany		East Germany	
Variable	Parameter	SE	Parameter	SE
Intercept	-1.2956	0.97	2.5444	1.0417
$HIGH_{96}$	-0.0022	0.0038	-0.0035	0.0049
R&D	0.7027	0.5435	7.2354	9631.206
COOP	-0.2519	0.6203	-6.724	9631.206
MR	-0.029	0.2556	-0.5718	0.3037
$STATE_{96}$	-0.2126	0.1164	-0.3545	0.1238
COMP	0.1787	0.1363	-0.0608	0.1542
PROF	-0.2244	0.3001	-0.276	0.379
ONEMAN	5.2694	10611.47	0.723	1.2341
PART	1.0705	0.5299	-0.5138	0.6674
LIMTD	0.8225	0.4423	0.0304	0.4181
CAPT	0.6959	0.5036	-0.7434	0.5792
PUBL	-0.0037	0.4875	-0.5207	0.7955
UNION	0.0194	0.3406	-0.3129	0.3002
INV_{97}	0.0000	0.0000	-0.0000	0.0000
ICT_{96}	0.1513	0.2272	-0.6361	0.2481
DEV_{96}	0.0287	0.0992	-0.0613	0.1032
ORGA	0.6562	0.406	1.199	0.4354
$SIZE_{50}$	-0.179	0.5683	-0.1855	0.546
$SIZE_{500}$	0.1408	0.3021	0.0428	0.388
	LR-Value		LR-Value	
Industry	46.14		37.28	
LR-Test	191.51		124.60	

Table F.6: Estimation Results for Organizational Changes

Variable	West Germany Parameter	West Germany SE	East Germany Parameter	East Germany SE
Intercept	-2.4812	0.8418	0.5447	1.0268
$HIGH_{96}$	0.0106	0.0034	-0.0007	0.0049
COMP	0.2347	0.1273	-0.3139	0.1791
ONEMAN	-7.6161	10611.46	0.5582	1.187
PART	-0.0151	0.4896	0.2126	0.6936
LIMTD	-0.0158	0.4374	0.391	0.5037
CAPT	0.1031	0.4462	0.9289	0.6089
PUBLIC	0.0249	0.472	7.1612	15012.68
UNION	0.5452	0.2635	0.104	0.2535
INV_{97}	0.0000	0.0000	-0.0000	0.0000
ICT_{96}	0.1202	0.1749	0.2199	0.1975
ICT_{97}	0.5451	0.221	0.1205	0.2339
DEV_{96}	-0.0446	0.0819	0.2565	0.1012
INNO	0.2133	0.1736	0.9125	0.2192
$SIZE_{50}$	-0.2861	0.5205	-0.0876	0.5223
$SIZE_{500}$	-0.4886	0.199	-0.0901	0.3128
	LR-Value		LR-Value	
Industry	12.05		16.20	
LR-Test	146.88		88.27	

Table F.7: Estimated Covariances between Labor Demand and Probit Equations

	West Germany		
	High Skilled	Medium Skilled	Low Skilled
Product Innovations	0	0.038	0.014
	(·)	(0.004)	(0.006)
Organizational Changes	0.42	0.006	0
	(0.08)	(0.03)	(·)
	West Germany		
	High Skilled	Medium Skilled	Low Skilled
Product Innovations	0.032	0	0.015
	(0.008)	(·)	(0.007)
Organizational Changes	0	0.013	0
	(·)	(0.018)	(·)

Note: Standard errors in parentheses.

Appendix G

Additional Material to Chapter 8

G.1 A Multiple Treatment Matching Protocol

The most simple matching protocol for the estimation of $\Delta^{ml|m} = E_i[Y_i(m) - Y_i(l)|D_i = m]$ includes the following steps:[1]

1. Specifying and estimating a multinomial choice model

As a result one obtains $\hat{p}^0(X_i), \hat{p}^1(X_i), \ldots, \hat{p}^M(X_i)$. The appropriate model depends on the problem at hand. If e.g. the treatments are inherently ordered, an ordered probit model applies whereas if there is no ordering, multinomial logit or probit models should be estimated. If one is only interested in $\Delta^{ml|m}$, an alternative is to estimate directly the conditional probabilities $p(D_i = l|D_i = m, l)$ using a binomial logit or probit model.

In order to improve the matching quality one can restrict the sample to the common support. In this case for every combination of l and m with $l, m \in \{0, 1, 2, \ldots, M\}$ all observations in the treatment group $D_i = l$ are deleted where the probabilities $p^l(X_i)$ and $p^m(X_i)$ are larger than the smallest maximum and smaller than the largest minimum in the control group $D_i = m$. If one is merely interested in $\Delta^{ml|m}$ it is sufficient to restrict only this subgroup to the common support.[2]

2. Estimating the conditional expected outcome variables

a. Choose one individual out of the group of individuals which participated in treatment m and delete it from that pool. Calculate for this individual either $\{p(D_i = m|X_i), p(D_i = l|X_i)\}$ or directly $p(D_i = l|D_i = m, l)$.

---

[1] Cf. Lechner (2001, 2002, 2002b) and Gerfin and Lechner (2002).

[2] Cf. Gerfin and Lechner (2002), p. 868.

b. Find one individual in the sub-sample of participants in treatment l that is as close as possible to the chosen treated individual in terms of his probabilities $\{p(D_i = m|X_i), p(D_i = l|X_i)\}$ or in terms of $p(D_i = l|D_i = m, l)$. In the first case a Mahalanobis distance can be calculated and additionally other covariates included. Do not delete this individual, i.e. conduct a matching with replacement. The reason is that unlike in the binomial case where we typically have more non-treated individuals than treated, in the multiple treatment model each group will act as a treated as well as a control group. Therefore requiring the number of comparisons to be larger than the number of treated individuals makes no sense.

c. Repeat step a. and b. until no treated participant m is left.

d. The mean difference of the outcome-variables for both groups may serve as an estimate of $\Delta^{ml|m} = E_i[Y_i(m) - Y_i(l)|D_i = m]$.

e. Having obtained an estimate of $\Delta^{ml|m} = E_i[Y_i(m) - Y_i(l)|D_i = m]$ and noting that any other effect $\Delta^{mj|j} = E_i[Y_i(m) - Y_i(j)|D_i = j]$ with $j \neq m$ can be obtained analogously, all other treatment effects can be obtained using the following relations:

$$\begin{aligned}
\Delta^{ml|m,l} &= E_i[Y_i(m) - Y_i(l)|D_i = m, l] \\
&= p(D_i = m|D_i = m, l)E_i[Y_i(m) - Y_i(l)|D_i = m] \\
&\quad + (1 - p(D_i = m|D_i = m, l))\, E_i[Y_i(m) - Y_i(l)|D_i = l] \\
&= p(D_i = m|D_i = m, l)\Delta^{ml|m} + (1 - p(D_i = m|D_i = m, l))\, \Delta^{ml|l} \\
\Delta^{ml} &= \sum_{j=0}^{M} (E_i[Y(m)|D_i = j] - E_i[Y(l)|D_i = j])\, p(D_i = j) \\
&= \sum_{j=0}^{M} \Delta^{ml|j} p(D_i = j).
\end{aligned}$$

The variance of these estimators can be calculated according to the binomial case and additionally the same diagnostic checks can be conducted.

G.2 Figures

Figure G.1: Total Employment from 1993 to 1997

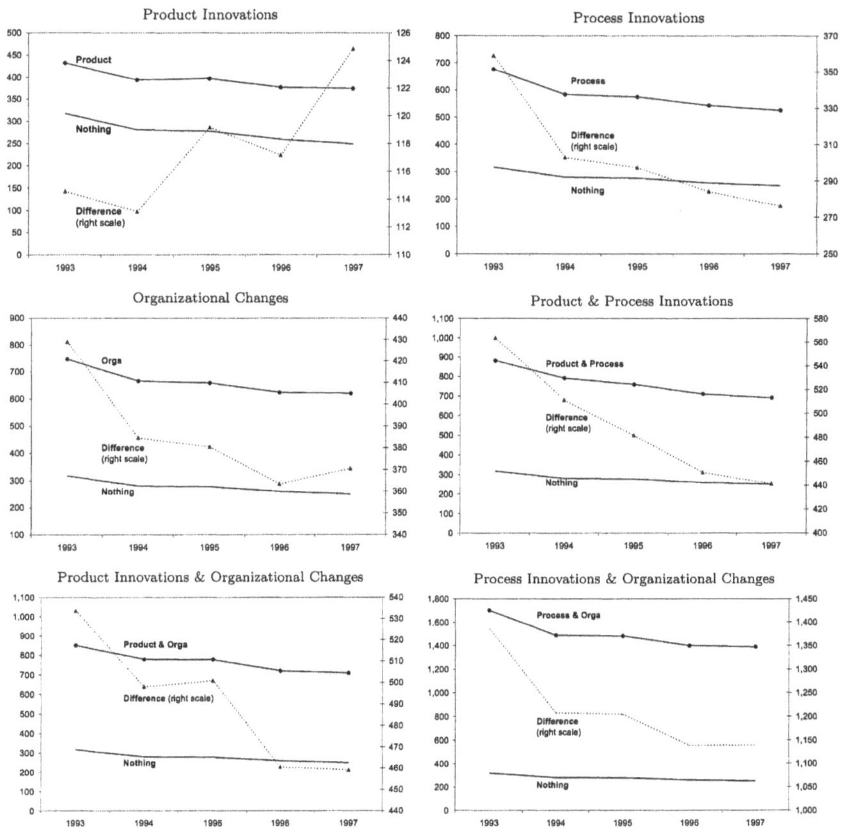

Figure G.2: Employment Shares for Low Skilled Employees from 1993 to 1997

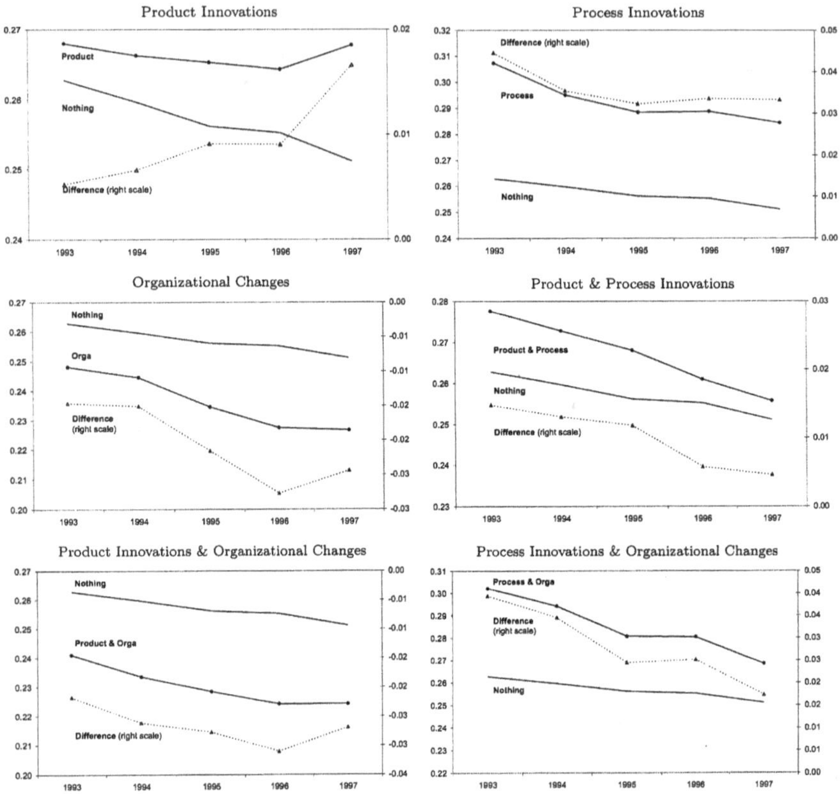

Figure G.3: Employment Shares for High Skilled Employees from 1993 to 1997

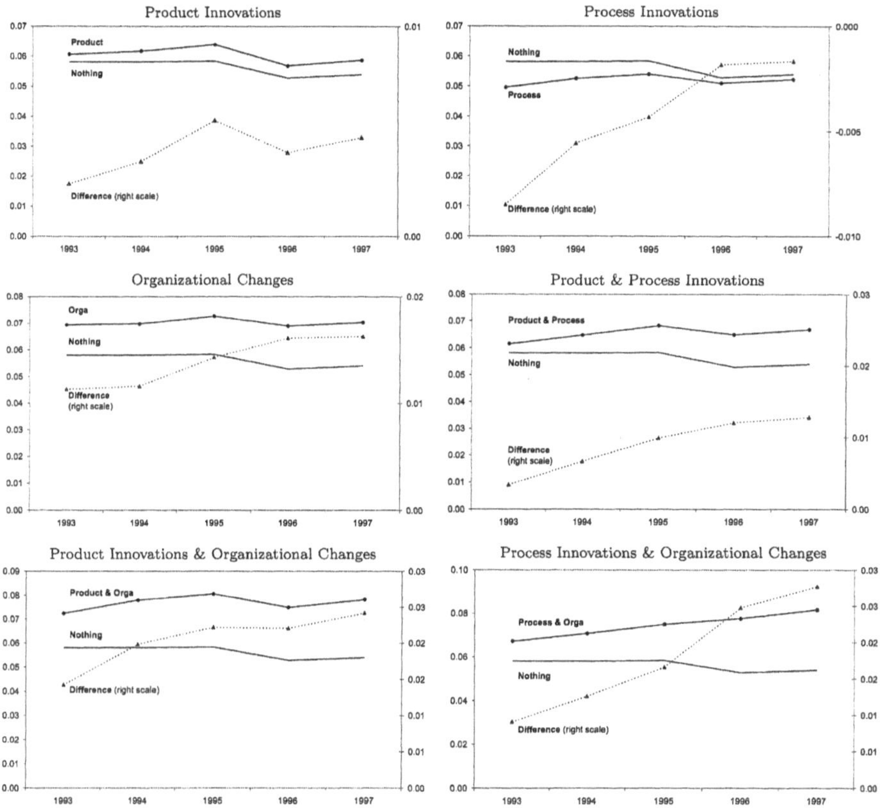

Figure G.4: Employment Shares for Young Employees from 1993 to 1997

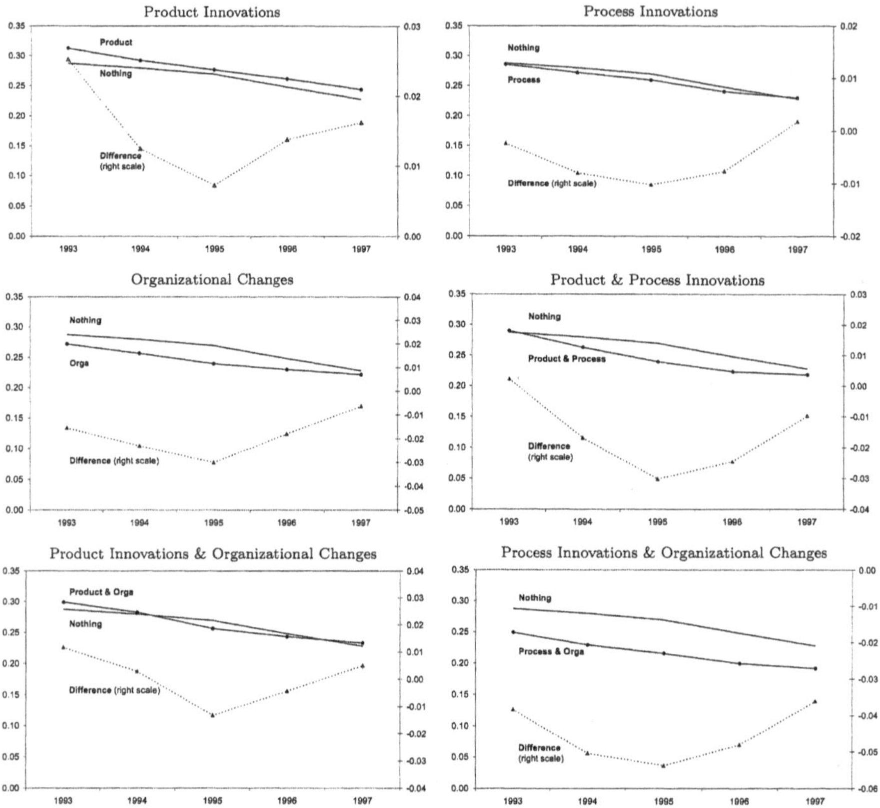

Figure G.5: Employment Shares for Elder Employees from 1993 to 1997

Product Innovations

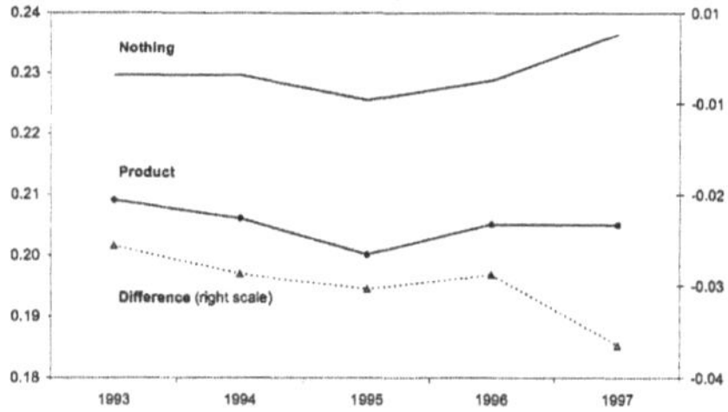

Process Innovations

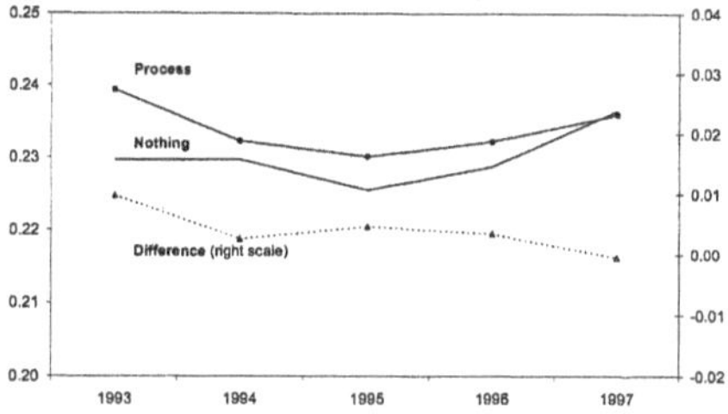

Organizational Changes

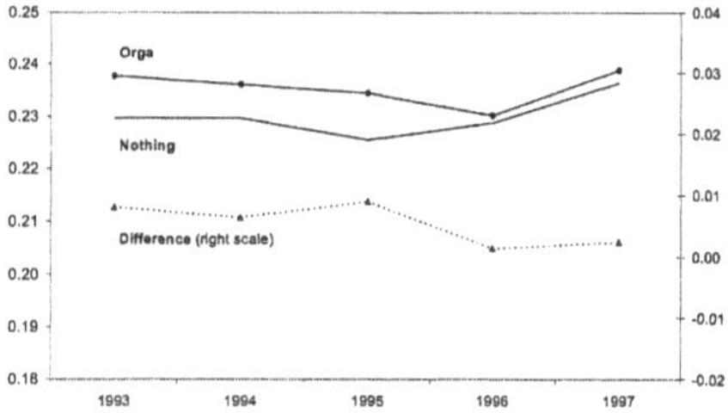

Product & Process Innovations

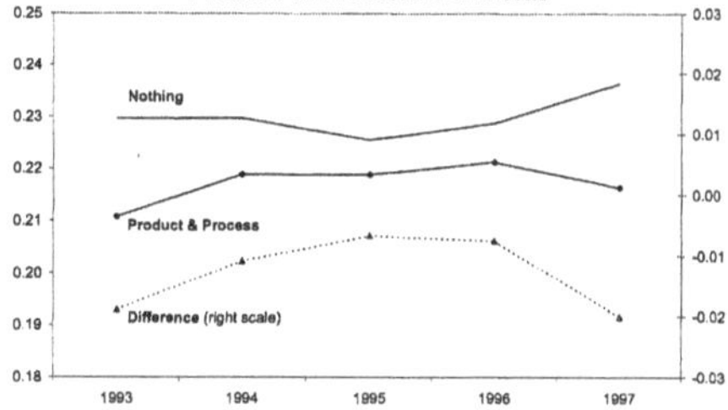

Product Innovations & Organizational Changes

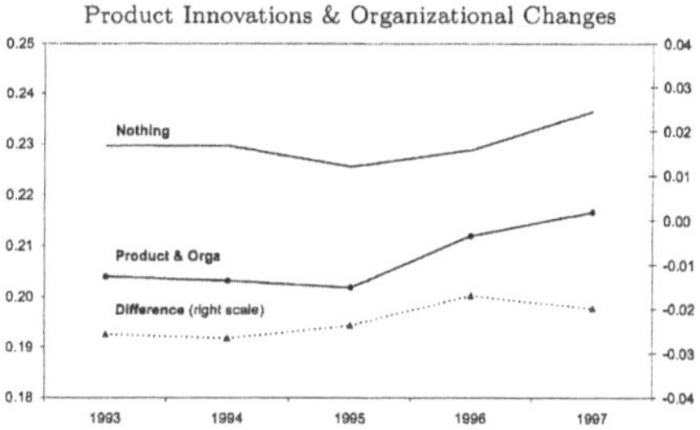

Process Innovations & Organizational Changes

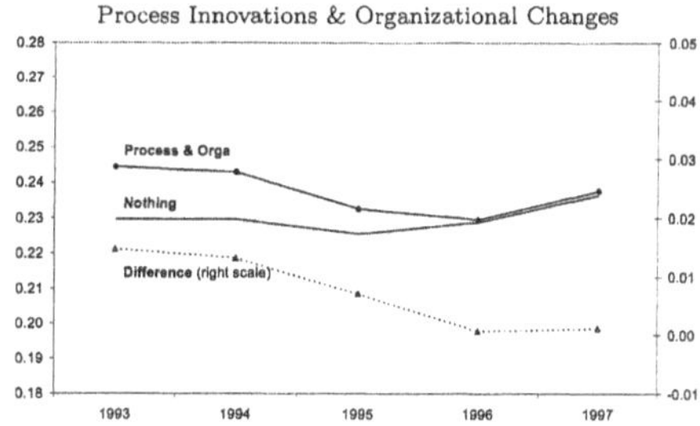

Abbreviations

CDF	Cumulative density function
DF	Degrees of freedom
DGP	Data generating process
PDF	Population density function
ATE	Average treatment effect
ATET	Average treatment effect on the treated
ATEU	Average treatment effect on the untreated
BAE	Before-after-estimator
Cf.	Confer (compare)
CIA	Conditional independence assumption
DID	Difference-in-differences
e.g.	Exempli gratia (for example)
FIML	Full information maximum likelihood
GHK	Geweke-Hajivassiliou-Keane (simulator)
GMM	Generalized method of moments
HRM	Human resource management
i.e.	Id est (that is to say)
IA	Independence assumption
IAB	Institut für Arbeitsmarkt und Berufsforschung (Institute for Employment Research)
ICT	Information and communication technology
IIA	Independence of irrelevant alternatives
LATE	Local average treatment effect
LIAB	Linked employer employee dataset
LR	Likelihood ratio
ML	Maximum likelihood
MND	Multivariate normal distribution
MNL	Multinomial logit model
MSEM	Mixed simultaneous equation model
ND	Normal distribution
OECD	Organisation for Economic Co-operation and Development
OLS	Ordinary least squares
R&D	Research and development
RMSE	Root mean squared error
SBTC	Skill biased technological change
SD	Standard deviation
SE	Standard error
SME	Small and medium sized establishments
SUTVA	Stable unit treatment value assumption

List of Tables

List of Figures

Bibliography

ABOWD, J. M., AND F. KRAMARZ (1999a): "Econometric Analyses of Linked Employer-Employee Data," *Labour Economics*, 6, 53–74.

——— (1999b): "The Analysis of Labor Markets Using Matched Employer-Employee Data," in *Handbook of Labor Economics*, ed. by O. Ashenfelter, and D. Card, vol. 3, pp. 2629–2710. Elsevier Science.

AKAIKE, H. (1973): "Information Theory and an Extension of the Likelihood Ratio Principle," in *Second International Symposium of Information Theory*, ed. by B. N. Petrov, and F. Csaki, pp. 267–281. Akademinai Kiado.

ALBERS, S., AND S. EGGERS (1991): "Organisatorische Gestaltung von Produktinnovations-Prozessen - Führt der Wechsel des Organisationsgrades zu Innovationserfolg?," *Schmalenbachs Zeitschrift für betriebswirtschaftliche Forschung*, 43(1), 44–65.

ALMUS, M., D. ENGEL, AND E. A. NERLINGER (1999): "Wachstumsdeterminanten junger Unternehmen in den alten und neuen Bundesländern: Ein Vergleich zwischen innovativen und nicht-innovativen Unternehmen," ZEW Discussion Paper No. 99-09, Centre for European Economic Research.

ALMUS, M., E. A. NERLINGER, AND F. STEIL (1999): "Growth Determinants of Start-Ups in Eastern Germany: A comparison between Innovative and Non-Innovative Firms," ZEW Discussion Paper No. 99-05, Centre for European Economic Research.

AMEMIYA, T. (1974): "Multivariate Regression and Simultaneous Equation Models when the Dependent Variables are Truncated Normal," *Econometrica*, 42(6), 999–1012.

——— (1979): "The Estimation of a Simultaneous-Equation Tobit-Model," *International Economic Review*, 20(1), 999–1012.

ANGRIST, J. D. (1991): "Instrumental Variable Estimation of Average Treatment Effects in Econometrics and Epidemiology," Discussion Paper, National Bureau of Economic Research.

——— (1995): "Conditioning on the Probability of Selection to Control Selection Bias," Technical Working Paper 181, National Bureau of Economic Research.

ANGRIST, J. D., AND N. EVANS (1990): "Children and Their Parents' Labor Supply: Evidence from Exogenous Variation in Family Size," *The American Economic Review*, 88, 450–477.

ANGRIST, J. D., AND G. W. IMBENS (1995): "Two-Stage Least Squares Estimation of Average Causal Effects in Models with Variable Treatment Intensity," *Journal of the American Statistical Association*, 90, 431–442.

ANGRIST, J. D., AND A. KRUEGER (1991): "Does Compulsory Schooling Affect Schooling and Earning?," *The Quarterly Journal of Economics*, 4, 979–1014.

ARNDS, P., AND H. BONIN (2002): "Frühverrentung in Deutschland: Ökonomische Anreize und institutionelle Strukturen," IZA Discussion Paper No. 666, Institute for the Study of Labor.

ASHENFELTER, O. (1978): "Estimating the Effect of Training Programs on Earnings," *The Review of Economics and Statistics*, 60(1), 47–57.

ASHENFELTER, O., AND D. CARD (1985): "Using the Longitudinal Structure of Earnings to Estimate the Effect of Training Programs on Earnings," *The Review of Economics and Statistics*, 67(4), 648–660.

ATHEY, S., AND S. STERN (1998): "An Empirical Framework for Testing Theories About Complementarity in Organizational Design," Discussion Paper.

AUTOR, D. H., L. F. KATZ, AND A. B. KRUEGER (1998): "Computing Inequality: Have Computers Changed the Labor Market?," *Quarterly Journal of Economics*, 113, 1169–1213.

BARTEL, A., AND N. SICHERMAN (1993): "Technological Change and Retirement Decisions of Older Workers," *Journal of Labor Economics*, 11, 162–183.

BARTEL, A. P., AND F. R. LICHTENBERG (1987): "The Comparative Advantage of Educated Workers in Implementing New Technology," *The Review of Economics and Statistics*, 69(1), 1–11.

BAUER, T. (2001): "Flexible Workplace Practices and Labor Productivity," IZA Discussion Paper No. 700, Institute for the Study of Labor.

BECKER, W., AND J. PETERS (2000): "University Knowledge and Innovation Activities," in *Technology and Knowledge: From the Firm to Innovation Systems*, ed. by P. Saviotti, and B. Noteboom, pp. 80–117.

BEISE, M., AND H. STAHL (1998): "Public Research and Industrial Innovations in Germany," ZEW Discussion Paper No. 98-37, Centre for European Economic Research.

BELL, B. D. (1996): "Skill-Biased Technical Change and Wages: Evidence from a Longitudinal Data Set," Discussion Paper, Institute of Economics & Statistics, University of Oxford.

BELLMANN, L. (1997): "The IAB Establishment Panel with an Exemplary Analysis of Employment Expectations," IAB Topics 20, Institute for Employment Research.

BELLMANN, L., S. BENDER, AND T. SCHANK (1999): "Flexibilität der Qualifikationsstruktur aus betrieblicher Sicht: Substitutionalität oder Komplementarität," *Jahrbücher für Nationalökonomie und Statistik*, 219(1+2), 109–125.

BELLMANN, L., M. CALIENDO, R. HUJER, AND D. RADIĆ (2002): "Beschäftigungswirkungen technisch-organisatorischen Wandels: Eine mikroökonometrische Analyse mit dem Linked IAB-Panel," *Mitteilungen aus der Arbeitsmarkt- und Berufsforschung*, 35, 506–522.

BELLMANN, L., AND T. SCHANK (2002): "Innovations, Wages and Demand for Heterogenous Labour: New Evidence from a Matched Employer-Employee Dataset," IZA Discussion Paper No. 112, Institute for the Study of Labor.

BENDER, S., A. HAAS, AND C. KLOSE (2000): "IAB Employment Subsample 1975-1995: Opportunities for Analysis Provided by the Anonymised Subsample," IZA Discussion Paper No. 117, Institute for the Study of Labor.

BERMAN, E., B. J., AND Z. GRILICHES (1994): "Changes in the Demand for Skilled Labor within U.S. Manufacturing: Evidence from the Annual Survey of Manufactures," *Quarterly Journal of Economics*, 109, 367–397.

BERMAN, E., J. BOUND, AND Z. GRILICHES (1994): "Changes in the Demand for Skilled Labor within U.S. Manufacturing: Evidence from the Annual Survey ," *Quarterly Journal of Economics*, pp. 367–397.

BERMAN, E., J. BOUND, AND S. MACHIN (1998): "Implications of Skill-Biased Technological Change: International Evidence," *Quarterly Journal of Economics*, 113, 1245–1279.

BERNDT, E. R. (1991): *The Practice of Econometrics: Classic and Contemporary.* Boston et. al.

BERTSCHEK, I., AND U. KAISER (2004): "Productivity Effects of Organizational Change: Microeconometric Evidence," *Management Science*, forthcoming.

BLACK, S., AND L. LYNCH (2001): "How to Compete: The Impact of Workplace Practices and Information Technology on Productivity," *The Review of Economics and Statistics*, 83(3), 434–445.

BLECHINGER, D., ET AL. (1998): "The Impact of Innovation on Employment in Europe: An Analysis using CIS Data," ZEW Discussion Paper No. 98-02, Centre for European Economic Research.

BLUNDELL, R., AND M. COSTA DIAS (2002): "Alternative Approaches to Evaluation in Empirical Microeconomics," CWP Working Paper No. 10-02, The Institute for Fiscal Studies.

BLUNDELL, R. W., AND R. J. SMITH (1993): "Simultaneous Microeconometric Models with Censored or Qualitative Dependent Variables," *Handbook of Statistics*, 11, 117–143.

——— (1994): "Coherency and Estimation in Simultaneous Models with Censored or Qualitative Dependent Variables," *Journal of Econometrics*, 64, 355–373.

BRANSTETTER, L., AND M. SAKAKIBARA (1998): "Japanese Research Consortia: A Microecononometrical Analysis of Industrial Policy," *Journal of Industrial Economics*, 46(2), 207–233.

BRESNAHAN, T. F., E. BRYNJOLFSSON, AND L. M. HITT (2002): "Information Technology, Workplace Organization, and the Demand for Skilled Labor: Firm-Level Evidence," *The Quarterly Journal of Economics*, 117(1), 339–377.

BROWNING, M. (1992): "Children and Household Economic Behavior," *Journal of Economic Literature*, 30, 1434–1475.

BÖRSCH-SUPAN, A. (2002): "Labor Market Effects of Population Aging," MEA Discussion Paper, Mannheim Research Institute for the Economics of Aging.

BUTLER, J. S., AND G. PICONE (1999): "Logically Inconsistent Limited Dependent Variables Models," in *Generalized Method of Moments Estimation*, ed. by L. Matyas, pp. 301–312.

CAMERON, A., AND P. TRIVEDI (2001): "Essentials of Count Data Regression," in *A Comparison to Theoretical Econometrics*, ed. by B. Baltagi.

CAMERON, A. C., AND P. K. TRIVEDI (1998): *Regression Analysis of Count Data.* Cambridge University Press.

CARD, D., AND J. E. DINARDO (2002): "Skill Biased Technological Change and Rising Wage Inequality: Some Problems and Puzzles," Working Paper No. 8769, National Bureau of Economic Research.

CAROLI, E., AND J. V. REENEN (2001): "Skill Biased Organizational Change? Evidence from a Panel of British and French Establishments," *The Quarterly Journal of Economics*, 116, 1449–1492.

CARSTENSEN, V. (2002): "The From-Tayloristic-toHolistic-Organization Model From an Empirical Perspective," Discussion Paper, University of Hannover.

CHENNELS, L., AND J. VAN REENEN (2002): "Technical Change and the Structure of Employment and Wages: A Survey of the Microeconometric Evidence," in *Productivity, Inequality, and the Digital Economy*, ed. by N. Greenan, Y. L'Horty, and J. Mairesse, pp. 175–223. Cambridge.

CLARKE, K. (2003a): "A Simple Distribution-Free Test for Nonnested Hypotheses," Discussion Paper, University of Rochester.

CLARKE, K. A. (2001): "Testing Nonnested Models of International Relations: Reevaluating Realism," *American Journal of Political Science*, 3, 724–744.

——— (2003b): "Nonparametric Model Discrimination in International Relations," *Journal of Conflict Resolution*, 47(1), 72–93.

COHEN, W. (1995a): "Empirical Studies of Innovative Activity," in *Handbook of the Economics of Innovation and Technological Change*, ed. by P. Stoneman, pp. 182–264. Oxford and Cambridge.

COHEN, W. (1995b): "Empirical Studies of Innovative Activity," *Handbook of the Economics of Innovation and Technological Change*, pp. 182–264.

COHEN, W. M., AND S. KLEPPER (1992): "The Anatomy of Industry R&D Intensity Distributions," *The American Economic Review*, 82, 773–788.

COHEN, W. M., AND R. C. LEVIN (1989): "Empirical Studies of Market Structure," in *Handbook of Industrial Organization*, ed. by R. Schmalensee, and R. Willig, pp. 1059–1107.

COHEN, W. M., R. C. LEVIN, AND D. C. MOWERY (1987): "Firm Size and R&D Intensity: A Re-Examination," *Journal of Industrial Economics*, 35, 543–563.

COMMISSION OF THE EUROPEAN COMMUNITIES (2001): *2001 Innovation Scoreboard.*

CREPON, B., AND E. DUGUET (1997): "Research and Development, Competition and Innovation: Pseudo-Maximum Likelihood and Simulated Maximum Likelihood Methods Applied to Count Data Models with Heterogeneity," *Journal of Econometrics*, 79, 355–378.

CREPON, B., E. DUGUET, AND J. MAIRESSE (1998): "Research, Innovation, and Productivity: An Econometric Analysis at the Firm Level," Working Paper No. 6696, National Bureau of Economic Research.

CZARNITZKI, D. (2001): "Die Auswirkungen der Forschungs- und Technologiepolitik auf die Innovationsaktivitäten ostdeutscher Unternehmen," *Schmollers Jahrbuch*, 121(4), 539–560.

CZARNITZKI, D., AND A. FIER (2002): "Substitutive or Complementary? Innovation Subsidies in the German Service Sector," ZEW Discussion Paper No. 02-04, Centre for European Economic Research.

DAVID, P., B. HALL, AND A. TOOLE (2000): "Is Public R&D a Complement or Substitute for Private R&D? A Review of the Econometric Evidence," *Research Policy*, 29, 497–529.

DAVIDSON, R., AND J. G. MACKINNON (1981): "Several Tests for Model Specification in the Presence of Alternative Hypotheses," *Econometrica*, 49(3), 781–793.

DOERINGER, P., C. EVANS-KLOCK, AND D. TERKLA (1998): "Hybrids or Hodgepodges? Workplace Practices of Japanese and Domestic Startups in the United States," *Industrial and Labor Relations Review*, 51(2), 171–186.

DOLTON, P., AND D. O'NEILL (1996): "Unemployment Duration and the RESTART Effect: Some Experimental Evidence," *The Economic Journal*, 106(435), 387–400.

EBLING, G., AND N. JANZ (1999): "Export and Innovation Activities in the German Service Sector: Empirical Evidence at the Firm Level," ZEW Discussion Paper No. 99-53, Centre for European Economic Research.

ENGELN, J., ET AL. (2003): "Indikatoren zur Ausbildung im Hochschulbereich," Studien zum Innovationssytem Deutschlands, Centre for European Economic Research.

ENTORF, H., AND W. POHLMEIR (1990): "Employment, Innovation and Export Activity," in *Microeconometrics: Surveys and Applications*, ed. by J. Florens, et al., pp. 394–415.

ERNST, H. (2003): "Unternehmenskultur und Innovationserfolg - Eine empirische Analyse," *Schmalenbachs Zeitschrift für betriebswirtschaftliche Forschung*, 55(1), 23–44.

EVANS, W. N., AND R. M. SCHWAB (1995): "Finishing High School and Starting College: Do Catholic Schools Make a Difference?," *Quarterly Journal of Economics*, 110, 941–974.

FALK, A., AND E. FEHR (2003): "Why Labour Market Experiments?," *Labour Economics*, 10(4), 399–406.

FALK, A., R. LALIVE, AND J. ZWEIMÜLLER (2002): "The Success of Job Applications: A New Experimental Design to Assess the Impact," Discussion Paper, University of Zurich.

FALK, M. (1999): "Technological Innovations and the Expected Demand for Skilled Labour at the Firm Level," ZEW Discussion Paper No. 99-59, Centre for Economic Research.

FALK, M., AND K. SEIM (2001): "Impact of Information Technology on High-Skilled Labour in Services: Evidence from Firm Level Panel Data," *Economics of Innovation and New Technology*, 10, 289–323.

FARBER, H. S. (1996): "The Changing Face of Job Loss in the United States, 1981-1993," Working Paper No. 5596, National Bureau of Economic Research.

FELDER, J., G. LICHT, E. NERLINGER, AND H. STAHL (1996): "Factors Determining R&D and Innovation Expenditure in German Manufacturing Industry," in *Determinants of Innovation: The Message from New Indicators*, ed. by A. H. Kleinknecht, pp. 125–154.

FIER, A., AND D. HARHOFF (2002): "Die Evolution der bundesdeutschen Forschungs- und Technologiepolitik: Rückblick und Bestandsaufnahme," *Perspektiven der Wirtschaftspolitik*, 3(3), 279–301.

FISHER, R. A. (1971): *The Design of Experiments*. Hafner Publishing Company, New York.

FITZENBERGER, B. (1999): *Wages and Employment Across Skill Groups: An Analysis for West Germany*. Heidelberg.

FITZENBERGER, B., AND W. FRANZ (1998): "Flexibilität der Qualifikatorischen Lohnstruktur und Lastverteilung der Arbeitslosigkeit: Eine Ökonometrische Analyse für Westdeutschland," in *Zunehmende Ungleichheit? – Erklärungen und Konsequenzen*, ed. by B. Gahlen, et al., pp. 47–79.

FLAIG, G., AND H. ROTTMANN (1999): "Direkte und indirekte Beschäftigungseffekte von Innovationen - Eine empirische Paneldatenanalyse für Unternehmen des westdeutschen Verarbeitenden Gewerbes," *BeitrAB*, 229, 149–175.

FRANÇOIS, J., F. FAVRE, AND S. NEGASSI (2002): "Competence and Organization: Two Drivers of Innovation (A Micro-Econometric Study)," *Economics of Innovation and New Technology*, 11(3), 249–270.

FRANSES, P. H., AND R. PAAP (2001): *Quantitative Models in Marketing Research*. Cambridge.

FREEMAN, C., AND L. SOETE (eds.) (1997): *The Economics of Industrial Innovation.* London.

FRIEDBERG, L. (1999): "The Impact of Technological Change on Older Workers: Evidence from Data on Computers," Discussion Paper, University of California.

GALHARDI, R. M. (1999): "The Impact of Trade and Technology on the Skill Profile in Brazil and the Republic of Korea," ILO Working Paper No. 40, International Labour Office.

GERFIN, M., AND M. LECHNER (2002): "A Microeconometric Evaluation of the Active Labour Market Policy in Switzerland," *The Economic Journal*, 112, 855–893.

GEROSKI, P. A. (1990): "Innovation, Technological Opportunity, and the Market Structure," *Oxford Economic Papers*, 42, 586–602.

GEROSKI, P. A., AND C. F. WALTERS (1995): "Innovative Activities Over the Business Cycle," *Economic Journal*, 105, 916–928.

GOURIEROUX, C., A. MONTFORT, AND A. TROGNON (1984a): "Pseudo-Maximum Likelihood Methods: Applications to Poisson Models," *Econometrica*, 52(3), 701–720.

——— (1984b): "Pseudo-Maximum Likelihood Methods: Theory," *Econometrica*, 52(3), 681–700.

GREENE, W. H. (1996): "Marginal Effects in the Bivariate Probit Model," Working Paper EC-96-11, Department of Economics, Stern School of Business, New York University.

——— (2003): *Econometric Analysis.* Prentice Hall, New Jersey.

GRILICHES, Z. (1990): "Patent Statistics as Economic Indicators: A Survey," *Journal of Economic Literature*, 28, 1661–1707.

——— (1995): "R&D and Productivity: Econometric Results and Measurement Issues," in *Handbook of the Economics of Innovation and Technological Change*, ed. by P. Stoneman, pp. 52–89. Oxford and Cambridge.

GRILICHES, Z., AND H. REGEV (1999): "R&D, Government Support, and Firm Productivity in Israeli Industry, 1975-1994," Working Paper, National Bureau of Economic Research.

GRUPP, H. (1997): *Messung und Erklärung des Technischen Wandels.* Berlin et. al.

HALL, B. H., Z. GRILICHES, AND J. H. HAUSMAN (1986): "Patents and R&D: Is There A Lag?," *International Economic Review*, 27(2), 265–284.

HALTIWANGER, J., ET AL. (eds.) (1999): *The Creation and Analysis of Employer-Employee Matched Data.* North-Holland.

HAMERMESH, D. S. (1986): "The Demand for Labor in the Long Run," in *Handbook of Labor Economics*, ed. by O. Ashenfelter. Elsevier Science, Amsterdam.

——— (1993): *Labor Demand.* Princeton University Press.

HARHOFF, D. (1998): "Are There Financing Constraints for Innovation and Investment in German Manufacturing Firms?," *Annales d'Économie et de Statistique*, 49/50, 421–456.

HARTMAN, R. S. (1991): "A Monte Carlo Analysis of Alternative Estimators in Models Involving Selectivity," *Journal of Business & Economic Statistics*, 9(1), 41–49.

HARVARD BUSINESS ESSENTIALS (2003): *Managing Creativity and Innovation.* Boston.

HAUSMAN, J., B. HALL, AND Z. GRILICHES (1984): "Econometric Models of Count Data with an Application of the Patents R&D Relationship," *Econometrica*, 52(4), 909–938.

HAUSMAN, J., AND D. MCFADDEN (1984): "Specification Tests for the Multinomial Logit Model," *Econometrica*, 52(5), 1219–1240.

HECKMAN, J., H. ICHIMURA, AND P. TODD (1997): "Matching as an Econometric Evaluation Estimator," *Review of Economic Studies*, 64, 605–654.

HECKMAN, J., R. LALONDE, AND J. SMITH (1999): ""The Economics and Econometrics of Active Labor Market Programs"," in *Handbook of Labor Economics*, ed. by O. Ashenfelter, and D. Card, vol. 3, pp. 1865–2097. Amsterdam: Elsevier.

HECKMAN, J., AND J. SMITH (1995): "Assessing the Case for Social Experiments," *Journal of Economic Perspectives*, 9, 85–110.

HERBERTSON, T. T., AND J. M. ORSZAG (2003): "The Early Retirement Burden: Assessing the Costs of Continued Prevalence of Early Retirement in OECD Countries," Discussion Paper, University of Iceland.

HITT, L., AND E. SNIR (1999): "The Role of Information Technology in Modern Production: Complement or Substitute to Other Inputs," Working Paper, University of Pennsylvania, Wharton School.

HOLLAND, P. (1986): "Statistics and Causal Inference," *Journal of American Statistical Association*, 81, 945–970.

HOLMSTROM, B., AND P. MILGROM (1994): "The Firm as an Incentive System," *The American Economic Review*, 84(4), 972–9991.

HOROWITZ, I. (1962): "Firm Size and Research Activity," *Southern Economic Journal*, 28, 298–301.

HUJER, R., AND M. CALIENDO (2001): "Evaluation of Active Labour Market Policy - Methodological Concepts and Empirical Estimates," in *Soziale Sicherung in einer dynamischen Gesellschaft*, ed. by I. Becker, N. Ott, and G. Rolf, pp. 583–617. Frankfurt/M.

HUJER, R., M. CALIENDO, AND D. RADIĆ (2001): "Estimating the Effects of Wage Subsidies on the Labour Demand in West Germany Using the IAB Establishment Panel," *ifo Studien*, 47, 163–197.

——— (2002): "Skill Biased Technological and Organizational Change: Estimating a Mixed Simultaneous Equation Model Using the IAB Establishment Panel," IZA Discussion Paper No. 566, Institute for the Study of Labor.

ICHINOWSKI, C., T. KOCHAN, D. LEVINE, C. OLSON, AND G. STRAUSS (1996): "What Works at Work: Overview and Assessment," *Industrial Relations*, 35(3), 299–333.

ICHNIOWSKI, C., K. SHAW, AND G. PRENNUSHI (1997): "The Effects of Human Resource Management Practices on Productivity: A Study of Steel Finishing Lines," *The American Economic Review*, 87(3), 291–313.

IMBENS, G. W. (2000): "The Role of the Propensity Score in Estimating Dose-Response Functions," *Biometrika*, 87, 706–710.

IRWIN, D. A., AND P. J. KLENOW (1996): "High-Tech R&D Subsidies: Estimating the Effects of Sematech," *Journal of International Economics*, 40, 232–344.

JANZ, N., G. EBLING, S. GOTTSCHALK, B. PETERS, AND T. SCHMIDT (2002): "Innovation Activities in the German Economy," Report on Indicators from the Innovation Survey 2001, Centre for European Economic Research.

JANZ, N., AND G. LICHT (eds.) (2003): *Innovationsforschung heute: Die Mannheimer Innovationspanels*, vol. 63 of *ZEW-Wirtschaftsanalysen*. Baden-Baden.

JONES, C. I., AND J. C. WILLIAMS (1998): "Measuring the Social Return to R&D," *The Quarterly Journal of Economics*, 113, 1119–1135.

JUHN, C. (1992): "Decline of Male Labor Market Participation: The Role of Declining Market Opportunities," *The Quarterly Journal of Economics*, 107, 79–121.

KAISER, U. (2000): "New Technologies and the Demand for Heterogeneous Labor: Firm-level Evidence for German Business-Related Services," *Economics of Innovation and New Technology*, 9, 465–484.

KAMIEN, M. I., AND SCHWARTZ (1985): "Market Structure and Innovation: A Survey," *Journal of Economic Literature*, 13, 1–37.

KATSOULACOS, Y. S. (1986): *The Employment Effect of Technical Change*. University of Nebraska Press.

KATZ, L. F., AND K. M. MURPHY (1992): "Changes in Relative Wages, 1963-1987: Supply and Demand Factors," *Quarterly Journal of Economics*, 107(1), 35–78.

KIM, E. H., AND V. SINGAL (1993): "Mergers and Market Power: Evidence from the Airline Industry," *The American Economic Review*, 83(3), 549–569.

KLETTE, T., AND J. MØEN (1999): "From Growth Theory to Technology Policy - Coordination Problems in Theory and Practice," *Nordic Journal of Political Economy*, 25, 53–74.

KLETTE, T. J. AND MØEN, J. AND GRILICHES, Z. (2000): "Do Subsidies to Commercial R&D Reduce Market Failures? Microeconometric Evaluation Studies," *Research Policy*, 29, 471–495.

KLINE, S., AND N. ROSENBERG (1986): "An Overview of Innovation," in *The Positive Sum Strategy: Harnessing Technology for Economic Growth*, ed. by R. Landau, and N. Rosenberg, pp. 275–306.

KÖNIG, H., H. BUSCHER, AND G. LICHT (1995): "Employment, Investment and Innovation at the Firm Level," in *OECD Jobs Study: Evidence and Explanations*. Organisation for Economic Co-operation and Development.

KÖLLING, A. (2000): "The IAB-Establishment Panel," *Schmollers Jahrbuch*, 120, 291–300.

KRAFT, K. (1989): "Market Structure, Firm Characteristics and Innovative Activity," *Journal of Industrial Economics*, 37, 329–336.

KRUEGER, A., AND D. CARD (1994): "Minimum Wages and Employment: A Case Study of the Fast Food Industry in New Jersey and Pennsylvania," *The American Economic Review*, 84(4), 772–793.

KRUEGER, A. B. (1993): "How Computers Have Changed the Wage Structure: Evidence From Microdata, 1984-1989," *Quarterly Journal of Economics*, 108(1), 33–60.

KRUGMAN, P. (1994): "Past and Prospective Causes of High Unemployment," in *Reducing Unemployment: Current Issues and Policy Options*. Federal Reserve Bank of Kansas City.

LALONDE, R. (1986): "Evaluating the Econometric Evaluations of Training Programs with Experimental Data," *The American Economic Review*, 76(4), 6094–620.

LAMBERT, D. (1992): "Zero-Inflated Poisson Regression, With an Application to Defects in Manufacturing," *Technometrics*, 34(1), 1–14.

LECHNER, M. (2000): "A Note on the Common Support Problem in Applied Evaluation Studies," Discussion Paper, University of St. Gallen.

——— (2001): "Identification and Estimation of Causal Effects of Multiple Treatments Under the Conditional Independence Assumption," in *Econometric Evaluation of Labour Market Policies*, ed. by M. Lechner, and F. Pfeiffer, pp. 43–58.

——— (2002a): "Program Heterogeneity and Propensity Score Matching: An Application to the Evaluation of Active Labor Market Policies," *The Review of Economics and Statistics*, 84, 205–220.

——— (2002b): "Some Practical Issues in the Evaluation of Heterogeneous Labour Market Programmes by Matching Methods," *Journal of the Royal Statistical Society. Series A (General)*, 165, 59–82.

LEE, L.-F. (1978): "Unionism and Wage Rates: A Simultaneous Equations Model with Qualitative and Limited Dependent Variables," *International Economic Review*, 19(2), 415–433.

——— (1981): "Simultaneous Equations Models with Discrete Endogenous Variables," in *Econometric Analysis of Discrete Data*, ed. by C. F. Manski, and D. McFadden, pp. 346–364. Cambridge.

LEE, S., R. FLORIDA, AND G. J. GATES (2002): "Innovation, Human Capital, and Creativity," Working Paper, Carnegie Mellon University and The Urban Institute.

LERNER, J. (1998): "The Government as Venture Capitalist: The Long-Run Impact of the SBIR Program," *Journal of Business*, 72, 285–318.

LEVIN, R., ET AL. (1987): "Appropriating the Returns from Industrial Research and Development," Discussion Paper No. 3, Brookings Papers on Economic Activity.

LI, K. (1998): "Bayesian Inference in a Simultaneous Equation Model with Limited Dependent Variables," *Journal of Econometrics*, 85, 387–400.

LINDBECK, A., AND D. J. SNOWER (1996): "Reorganization of Firms and Labor Market Inequality," *The American Economic Review*, 86(2), 315–321.

——— (2000): "Multitask Learning and the Reorganization of Work: From Tayloristic to Holistic Organization," *Journal of Labor Economics*, 18(3), 353–376.

LIPSEY, R. G. (2001): "Economic Growth, Technological Change and Economic Policy," Working Paper, Simon Fraser University.

LOPEZ-ACEVEDO, G. (2002): "Technology and Skill Demand in Mexico," Discussion Paper, World Bank.

LUNDVALL, B. A., AND F. S. KRISTENSEN (1997): "Organizational Change, Innovation and Human Ressource Development as a Response to Increased Competition," DRUID Working Paper No. 97-16, Danish Research Unity for Industrial Dynamics.

MACHIN, S., AND S. WADHWANI (1991): "The Effects of Unions on Organisational Change and Employment," *The Economic Journal*, 101(407), 835–854.

MADDALA, G. S. (1983): *Limited-Dependent and Qualitative Variables in Econometrics.* Cambridge.

MANSFIELD, E. (1986): "Patents and Innovation: An Empirical Study," *Management Science*, 32, 173–181.

MARKUSEN, J. R., ET AL. (1995): *International Trade: Theory and Evidence.* New York et. al.

MCFADDEN, D., AND P. RUUD (1994): "Estimation by Simulation," *The Review of Economics and Statistics*, 76(4), 591–608.

MEYER, B. D. (1995): "Natural and Quasi-Experiments in Economics," *Journal of Business & Economic Statistics*, 13(2), 151–161.

MICHIE, J., AND M. SHEEHAN (2000): "HRM Practices, R&D Expenditure and Innovative Investment: Evidence from the UK's 1990 Workplace Industrial Relations Survey (WIRS)," *Industrial and Corporate Change*, 29, 471–495.

MILGROM, P., AND J. ROBERTS (1990): "The Economics of Modern Manufacturing: Technology, Strategy, and Organization," *The American Economic Review*, 80(3), 511–528.

MÜLLER, D. C. (1967): "The Firm's Decision Process: An Econometric Investigation," *Quarterly Journal of Economics*, 81, 58–87.

MULLAHY, J. (1986): "Specifications and Testing of Some Modified Count Data Models," *Journal of Econometrics*, 33, 341–365.

NEIDERT, A. (1998): "Neues Meldeverfahren zur Sozialversicherung ab 1999," *Deutsche Rentenversicherung*, 5, 315–330.

NELSON, F., AND L. OLSON (1978): "Specification and Estimation of a Simultaneous-Equation Model with Limited Dependent Variables," *International Economic Review*, 19(3), 695–709.

NELSON, R. R. (1959): "The Simple Economics of Basic Scientific Research," *Journal of Political Economy*, 67, 297–306.

OECD (ed.) (1993): *The Proposed Standard Practice for Surveys or Research and Experimental Development, Frascati Manual 1993.* Paris.

——— (ed.) (1997): *The Measurement of Scientific and Technological Activities, Proposed Guidelines for Collecting and Interpreting Technological Innovation Data, Oslo Manual 1992.* Paris.

OLSON, L. (1979): "Maximum Likelihood Estimation of the Polychoric Correlation Coefficient," *Psychometrika*, 44(4), 443–460.

PAQUÉ, K.-H. (1998): "Internationaler Handel, Technischer Fortschritt und Lohndifferenzierung," in *Verteilungsprobleme der Gegenwart: Diagnose und Therapie*, ed. by B. Gahlen, et al., pp. 123–131.

PATEL, P., AND K. PAVITT (1989): "Patterns of Technological Activity: Their Measurement and Interpretation," in *Handbook of Industrial Organization*, ed. by R. Schmalensee, and R. Willig, pp. 1059–1107.

PENZKOFER, H. (1994): *Wirtschaftliche Effekte von Innovations- und FuE-Aktivitäten im Zeitraum 1986 bis 1990- eine Empirische Analyse auf der Grundlage des IFO Innovationstestpanels.* Tübingen.

——— (2003): "Innovationsaktivität in der Industrie 2001/2002: Leichter Rückgang auf hohem Niveau," *ifo Schnelldienst*, 56(2), 24–29.

PETIT, P. (1995): "Employment and Technological Change," in *Handbook of the Economics of Innovation and Technological Change*, ed. by P. Stoneman, pp. 52–89. Oxford and Cambridge.

PIL, F., AND J. MACDUFFIE (1996): "The Adoption of High-Involvement Work Practices," *Industrial Relations*, 35(3), 299–334.

PLEWIS, I. (2002): "Modelling Impact Heterogeneity," *Journal of the Royal Statistical Society. Series A (General)*, 165, 31–38.

POHLMEIER, W. (1989): *Simultane Probit- und Tobitmodelle.* Berlin et. al.

POHLMEIER, W., AND V. ULRICH (1995): "An Econometric Model of the Two-Part Decisionmaking Process in the Demand for Health Care," *The Journal of Human Resources*, 30, 339–361.

PUHANI, P. A. (2003): "A Test of the 'Krugman Hypothesis' for the United States, Britain, and Western Germany," IZA Discussion Paper No. 764, Institute for the Study of Labor.

RANSOM, M. R. (1987): "An Empirical Model of Discrete and Continuous Choice in Family Labor Supply," *The Review of Economics and Statistics*, 69(3), 465–472.

REICH, R. (2000): *The Future of Success.* New York.

RIFKIN, J. (1995): *The End of Work.* New York.

RIVERS, D., AND Q. H. VUONG (1988): "Limited Information Estimators and Exogeneity Tests for Simultaneous Probit Models," *Journal of Econometrics*, 39, 347–366.

ROMER, P. M. (2000): "Should the Government Subsidize Supply or Demand in the Market for Scientists and Engineers?," Working Paper 7723, National Bureau of Economic Research.

ROSENBAUM, P. R., AND D. B. RUBIN (1985): "Constructing a Control Group using Multivariate Matched Sampling Methods that Incorporate the Propensity Score," *The American Statistician*, 39, 33–38.

ROSENZWEIG, M. R., AND K. I. WOLPIN (2000): "Natural "Natural Experiments" in Economics," *Journal of Economic Literature*, 38, 827–874.

ROSS, D. R., AND K. ZIMMERMANN (1993): "Evaluating Reported Determinants of Labour Demand," *Labour Economics*, 1, 71–84.

ROTHWELL, R., ET AL. (1974): "SAPPHO Updated-Project SAPPHO II," *Research Policy*, 3, 258–291.

ROTTMANN, H., AND M. RUSCHINSKI (1997): "Sind Innovationen beschäftigungswirksam?," IFO Discussion Paper No. 17-18, Ifo Institute for Economic Research.

ROY, A. (1951): "Some Thoughts on the Distribution of Earnings," *Oxford Economic Papers*, 3, 135–145.

RUBIN, D. (1974): "Estimating Causal Effects to Treatments in Randomised and Nonrandomised Studies," *Journal of Educational Psychology*, 66, 688–701.

——— (1977): "Assignment to Treatment Group on the Basis of a Covariate," *Journal of Educational Studies*, 2, 1–26.

——— (1979): "Using Multivariate Matched Sampling and Regression Adjustment to Control Bias in Observational Studies," *Journal of the American Statistical Association*, 74, 318–328.

——— (1980): "Comment on Basu, D. Randomization Analysis of Experimental Data: The Fisher Randomization Test," *Journal of the American Statistical Association*, 75, 591–593.

SCHERER, F. M. (1965): "Firm Size, Market Structure, Opportunity, and the Output of Patented Inventions," *The American Economic Review*, 55, 1097–1125.

——— (1967): "Market Structure and the Employment of Scientists and Engineers," *The American Economic Review*, 57, 524–531.

SCHMIDT, P. (1978): "Estimation of a Simultaneous Equations Model with Jointly Dependent Continous and Qualitative Variables: The Union-Earnings Question Revisted," *International Economic Review*, 19(2), 453–465.

——— (1990): "Constraints on the Parameters in Simultaneous Tobit and Probit Model," in *Structural Analysis of Discrete Data with Econometric Applications*. The MIT Press.

SCHMOOKLER, J. (1962): "Economic Sources of Inventive Activity," *Journal of Economic History*, 22, 1–10.

SCHMOOKLER, J. (1966): *Invention and Economic Growth*. Cambridge.

SCHOLZ, L., ET AL. (1989): *Innovation, Wachstum und Beschäftigung : Einzelwirtschaftliche, sektorale und intersektorale Innovationsaktivitäten und ihre Auswirkungen auf die Wirtschaft in den achtziger Jahren*. Berlin.

SCHWARZ, G. (1978): "Estimating the Dimension of a Model," *The Annals of Statistics*, 6(2), 461–464.

SHARMA, S. (1996): *Applied Multivariate Techniques*. New York.

SIANESI, B. (2003): "Differential Effects of Swedish Active Labour Market Programmes for Unemployed Adults during the 1990s," IFS Working Paper No. 01-25, The Institute for Fiscal Studies.

SILVERMAN, B. (1986): *Density Estimation for Statistics and Data Analysis*. London.

SMALL, K. A., AND C. HSIAO (1985): "Multinomial Logit Specification Tests," *International Economic Review*, 26, 619–627.

SMITH, J., AND P. TODD (2000): "Does Matching Overcome LaLonde's Critique of Nonexperimental Estimators?," Discussion Paper.

SMOLNY, W., AND T. SCHNEEWEIS (1999): "Innovation, Wachstum und Beschäftigung," *Jahrbücher für Nationalökonomie und Statistik*, 218(3+4), 453–472.

SNOWER, D. J. (1998): "The Organizational Revolution and its Implications for Job Creation," in *The Role of Labor Market Institutions*, ed. by J. Gual, pp. 49–70. Cheltenham et. al.

SNOWER, D. J. (1999): "Causes of Changing Earnings Inequality," IZA Discussion Paper No. 29, Institute for the Study of Labor.

SOLOW, R. M. (1957): "Technical Change and the Aggregate Production Function," *The Review of Economics and Statistics*, 39(3), 312–320.

SPANOS, A. (1999): *Probability Theory and Statistical Inference*. Cambridge.

SPENCE, M. (1984): "Cost Reduction, Competition, and Industry Performance," *Econometrica*, 52, 101–121.

SPIEZIA, V., AND M. VIVARELLI (2002): "Innovation and Employment: A Critical Survey," in *Productivity, Inequality, and the Digital Economy*, ed. by N. Greenan, Y. L'Horty, and J. Mairesse, pp. 101–131. Cambridge.

STOLZENBERG, R. M., AND D. A. RELLES (1991): "Tools for Intuition about Sample Selection Bias and Its Correction," *American Sociological Review*, 62(3), 494–507.

TIROLE, J. (1988): *Theory of Industrial Organization.* MIT-Press.

VAN DEN BROEK, J. (1995): "A Score Test for Zero Inflation in a Poisson Distribution," *Biometrics*, 51, 738–743.

VELLA, F. (1998): "Estimating Models with Sample Selection Bias: A Survey," *The Journal of Human Resources*, 33(1), 127–169.

VEUGELERS, R. (1997): "Internal R&D Expenditures and External Technology Sourcing," *Research Policy*, 26, 303–315.

VILLARD, H. (1958): "Competition, Oligopoly and Research," *Journal of Political Economy*, 66, 483–497.

VUONG, Q. H. (1989): "Likelihood Ratio Tests for Model Selection and Non-Nested Hypotheses," *Econometrica*, 57(2), 307–333.

WALDMAN, D. M. (1981): "An Economic Interpretation of Parameter Constraints in a Simultaneous-Equations Model with Limited Dependent Variables," *International Economic Review*, 22(3), 730–739.

WALSH, V. (1984): "Invention and Innovation in the Chemical Industry: Demand-Pull or Discovery-Push," *Research Policy*, 13, 211–234.

WHITE, H. (1982): "Maximum Likelihood Estimation of Misspecified Models," *Econometrica*, 50(1), 1–26.

WILDE, J. (1999): *Gemischte simultane Modelle für Querschnittsdaten.* Berlin et. al.

WINKELMANN, R. (1997): *Econometric Analysis of Count Data.* Springer.

WINKELMANN, R. (2003): "Health Care Reform and the Number of Doctor Visits - An Econometric Analysis," Working Paper, University of Zurich.

WINSHIP, C., AND R. D. MARE (1992): "Models for Sample Selection Bias," *Annual Review of Sociology*, 18, 327–350.

WOLF, E., AND T. ZWICK (2002): "Reassessing the Impact of High Performance Workplaces," ZEW Discussion Paper No. 02-07, Centre for European Economic Research.

WOOD, A. (1995): "How Trade Hurt Unskilled Workers," *The Journal of Economic Perspectives*, 9(3), 57–80.

WOOLDRIDGE, J. M. (2001): *Econometric Analysis of Cross Section and Panel Data.* Cambridge.

ZELLNER, A. (1962): "An Efficient Method of Estimating Seemingly Unrelated Regressions and Tests for Aggregation Bias," *Journal of the American Statistical Association*, 57, 348–368.

ZIMMERMANN, K. F. (1987): "Innovation und Beschäftigung," in *Arbeitsmärkte und Beschäftigung: Fakten, Analysen, Perspektiven*, ed. by G. Bombach, et al., pp. 233–255. Tübingen.

——— (1991): "The Employment Consequences of Technological Advance: Demand and Labour Costs in 16 German Industries," *Empirical Economics*, 16, 253–266.

Kolibri 360 GmbH
Ferdinand-Jühlke-Straße 7
99095 Erfurt, Deutschland
produktsicherheit@kolibri360.de